Be Your Own HOUSE CONTRACTOR

How to Save 25 % Without Lifting a Hammer

Carl Heldmann

A Storey Publishing Book

Storey Communications, Inc.
Schoolhouse Road
Pownal, Vermont 05261

Edited by Deborah Balmuth
Cover design by Greg Imhoff
Text design by Cindy McFarland
Text production by Leslie Carlson
Indexed by Nan Badgett, Word•a•bil•i•ty

Printed in the United States by R.R. Donnelley
Second Edition, First Printing, December 1994

Library of Congress Cataloging-in-Publication Data

Heldmann, Carl.

Be your own house contractor: how to save 25 percent without lifting a hammer / Carl Heldmann. 3rd edition

p. cm.

"A Storey Publishing book."

Includes index

ISBN 0-88266-266-X (pbk.) :$12.95

1. House construction — Amateurs' manuals.
2. Contracting out — Amateurs' manuals. I. Title.

TH4815.H44 1995 94-36121
690'.837 — dc20 CIP

Contents

Preface to the Third Edition

The most expensive words in real estate are "Let's wait." If you wait for interest rates to fall before building a house, you can be sure that the costs of building will increase in the meantime. If you wait for building costs to decrease, it will never happen. **Now** is always the best time to build a house.

But what if you just can't afford "Now"? What do you do? Back in the early 1970s, my wife and I were faced with this very dilemma. We just couldn't afford to build our "dream house" and didn't know what to do. A friend suggested that by building our own home we would save enough money to make the project feasible. I thought he was crazy. Build my own house? I said. I don't know anything about building houses. "Relax," he said, and proceeded to explain to me that by merely acting as the general contractor and hiring all the expert subcontractors (tradespeople), I could save 25 percent or more on the cost of my dream home.

I was skeptical, I admit, but, being the type of person who enjoys finding better ways to solve problems, I decided to try contracting. And the rest, as they say, is history.

I built that first house, saved more than 25 percent, and went on to home building as a career, building millions of dollars worth of homes. Now I am involved in financing new construction. Along the way I wrote this book so that people such as yourself won't have to make as many mistakes on their first house as I did. I got through the process of building my first house, and I'm certain you will as well.

By using this book as a guide, you can avoid many of the pitfalls that plague general contractors, even experienced ones. You will quickly realize that you can not only save 25 percent or more, but by being the "boss" you'll get exactly the house you want, done the way you want it done. If you want to save a little more money, you can even do some of the labor yourself. But, more on that later.

This is not a technical book on how to perform the tasks of the various subcontractors you will be hiring. There are plenty of books and videos on carpentry, plumbing, and other specific skilled tasks if you decide you want to get that involved. This is a book that shows you how to be your own general contractor. You will see that you don't have to do any physical labor to save 25 percent or more. You don't even have to lift a hammer. You can hire the pros to do that.

I hope that the prospect of being able to save money now, as you build, will enable you to say, "Let's not wait any longer, let's build it **now.**" Who knows? Perhaps this venture will launch a new career for you. It certainly did for me. Every professional building general contractor started with his or her first home. Most of us didn't have a book to guide us. You do. Good Luck!

Acknowledgments

For their tremendous help, I thank my wife, Jane Prante Heldmann, Sharon W. Curran, and Joyce F. Carpenter. My special thanks to Clella Hunt Prante for her invaluable assistance.

Introduction

People always seem to think things are harder than they are. I don't know why this is so, but I'm the same way. Perhaps we fear what we don't know or maybe it's because someone told us "It's too hard." Then, when we finally get around to doing it, we discover that the task wasn't as hard as we anticipated.

Everyone who hasn't built a house before thinks it's too hard. For decades, I've been spreading the word that it isn't hard. But I've found that people tend to be skeptical of my enthusiastic belief that "you can do it." However, I've taught thousands of people over the years to do exactly what you are thinking of doing — be your own house contractor. They have come from all walks of life — housewives, salespeople, truck drivers, lawyers, doctors, you name it — and were all motivated by more or less the same reasons. The only variable is the size and style of the house that each one envisioned. And, most important, they all found that they *could* "do it," and that my enthusiasm was well founded.

Can *you* do it? Is being your own house contractor as easy as I'm making it out to be? Well, I'll tell you a true story that pretty well sums up my answer.

Not too many years ago, one morning at 8:00 I received a phone call from a woman I didn't know. Our conversation went something like this.

"Is this Carl Heldmann?"

"Yes, it is."

"Are you the one who says it's easy to build your own home?"

By now I was a little nervous and expected to hear the worst. But I said, "I sure am!"

"Well," she replied, "I just wanted to tell you that it's even easier than you said it was!"

I breathed an audible sigh of relief.

The woman went on to say, "My husband and I just finished building our own home in Windy Hill, and we saved so much money and had so much fun, that we bought the lot next door and we're going to do it again!"

What more could I possibly add to that testimonial?

You will notice that some technical terms are set in *italics* in this book. These are terms you should understand. If you don't, look them up in the Glossary (pages 91–94), where they are defined.

Throughout the book, you will find several sample contracts I have provided to acquaint you with their format. I urge you *not* to use these forms unless you have reviewed them with your attorney. State laws differ on what is required in such contracts; your attorney will know what changes may be necessary.

Chapter 1

Be Your Own General Contractor and Save

THIS IS A BOOK that will teach you how to be a general contractor for your own house. There's one big reason for doing it — to save money. How much you will save will vary considerably depending on local prices for labor and materials, land costs, the size of the house, and your ability to follow the steps outlined in this book.

The size of the house will be the largest determining factor, as most general contractors base their profit and overhead on a percentage of the total cost. A larger house will cost more, and therefore will include a larger builder's profit and overhead. Real estate commissions saved will also be greater as the size of the house increases.

If you do all that I suggest in this book, a goal of 25 percent savings is possible. If you think of how much you will have to earn — and pay taxes on — to get the money to build, the savings is even greater. Greater still is the fact that by reducing the amount of money you will need to borrow and then pay back with interest, you will save even more money.

Let's look at what you can expect to save.

Say that I built a house and I am offering it for sale to the general public. If I set a selling price of $100,000, typically, the land would have cost me $25,000, labor and materials $50,000, and my gross profit would have been $25,000.

Wow! you say. That's a lot of money to make off one product. Well, if it were that simple, you would be right. But before you get outraged with the building industry in general, let me show you where that gross profit goes when you build professionally.

First, I have to pay sales expense out of my gross profit. If that involves a real estate broker, it may cost as much as $7,000 (7 percent of $100,000). Next, like any business, I have business expense overhead. This varies with each builder, but the National Association of Home Builders suggests that home builders allocate 50 percent of their adjusted gross profit (after sales expense) for overhead expenses. These include, but are not limited to, phone, insurance, secretarial, gas, truck, car, rent, and office equipment. That leaves me with $9,000, which is half the adjusted gross of $18,000 ($25,000 – $7,000 real estate commission). That is my real expected net (before taxes) profit. You can see that when builders say they make less than 10 percent, they're not wrong.

But you aren't building professionally. You don't have sales expense. You don't have business expense overhead. You can take the entire gross profit of $25,000 and consider it yours. You may never even have to pay taxes on that gross profit if you follow the Internal Revenue Service guidelines for reinvesting your primary residence capital gains. Imagine! You could actually "make" this kind of money while you go about your regular walk in life.

Actual dollar amounts will vary depending on the time and place, but the percentage of actual "fair market value" (what a house should sell for) you can gross will remain fairly true. Does that mean you can expect to save $50,000 on a house that would sell for $200,000? You bet!

What You Need to Know

You need to know very little about building to be a general contractor. You don't need to have the technical knowledge about framing or bricklaying or wiring. Your subcontractors will know their business just as mine do. I'll help you make sure of that.

You may wish to pick up some information on various aspects of building, and that's fine. In most bookstores there are excellent "how-to" books for the do-it-yourselfer on almost all phases of construction. You may want them. But there's no way you can become a master of all trades. The role of the general contractor — which you will be — is as an organizer, a manager, not a tradesman. Your role is to get the job done — by other people.

Hire an Attorney

If, in any of the planning stages of building outlined in this book, you feel inadequate, hire a good real estate attorney. You will need the help of an attorney in certain steps anyway, such as in the closing of the loan. If you hire one now, the fee will be money well spent if it enables you to build your own home, and that fee is an actual cost of construction. Many attorneys will forgo hourly rates, or at least reduce them, if they are going to be involved throughout your project. Be sure you like the attorney in your first meeting. The two of you will have a long relationship, for the six or more months the house is under construction, so this is important. See page 58 for information on obtaining an attorney.

If you can balance a checkbook, read, and deal with people in a fair manner, you can build your own house. Don't look at building a house as one huge job. I don't do this after many years in the business. View each phase or step as a separate job and in this way the overall task will seem less monumental.

The most difficult things you will have to do will be behind you when you actually start construction. Sounds unbelievable, doesn't it? But it's true. Your job of planning and organizing will be 90 percent complete when you break ground . . . or it should be. At that point it is up to your team of experts — your subcontractors (subs) — to do its job. At least 99 percent of them will do their jobs correctly, even without you being at the site.

As you go along never be afraid to ask anyone a question about anything. Pride is foolish when it prevents you from asking a learning question.

Carpentry Crew

A definite help to you in building your own home, or even in helping you to make the decision to do so, is finding a good carpentry crew. The carpenter, or crew, is the key subcontractor. Having a good one before you start will help you make many intelligent decisions and will help your local lender to make a decision to lend you money. More on this later in this chapter.

A personal visit to a couple of building supply houses (usually the smaller ones) to ask for recommendations will give you more than enough leads to find a good carpentry crew.

If the first carpenter you find is too busy, ask him to recommend another. But usually another house can be worked into the first carpenter's scheduling. Your carpenter will be one of your best sources for finding many of your other subcontractors.

After selecting a good candidate for this job, check him out. He should give you a list of four or five jobs he's handled. Talk to the owners. Look at the work he did for them. Others to check with are banks and building suppliers.

Just a note now about this key subcontractor. Most good carpenters, generally the older ones — mid-thirties and up — have a very good knowledge of most phases of construction. They may be weak in their technical knowledge of electricity or mechanics, and may

Keeping the Records

Unless you're an exceptionally orderly person, you'll find yourself with construction papers at your desk at home, more in your car glove compartment, and others scattered in pockets or your place of business.

Avoid this by organizing yourself early. Get a briefcase, which will be your traveling office file. Buy some manila folders, a dozen at least. Then label them as you need them. You'll need one for finances, another for inspections, with a schedule of those inspections made out by you, one for each of the subs — and more.

There are two reasons for doing this. First, it will keep your work flowing more smoothly if you don't have to paw through papers to find what you want. Second, there's a feeling of confidence that you will get just from seeing your part of the job arranged in such a businesslike way.

not be good at finances or management, but most have built a house or houses completely. The good carpenters are easy to find, hard to get, and usually worth waiting for. They will also cost you a bit more, but in the long run, will save you money and time.

Time Involved

How much time is involved in being your own contractor? For each individual it will be different, but you'll spend more hours planning and getting ready than you will on the construction site — that is, if you let the subs do their jobs without your feeling that you must be there every minute.

In the planning stages you may spend a month of your spare time or several months, depending on how fast you are able to decide on all the various aspects that I will cover. For example, deciding on house plans can take an hour, a month, or even longer; the same thing is true with choosing land, specifications, and subcontractors.

The time involved after you start construction will be short. The maximum time ought not to exceed two hours a day; those hours need not interfere with your job or normal activities. You will have many helpers and free management. You can have real estate people (the ones you are dealing with on your lot) do some legwork by checking on restrictions, and getting information on septic systems and wells. Your suppliers can help you save time, too, by finding your subs for you, doing *"take-offs"* for estimating and ordering supplies, and giving technical advice.

Building your own home is an ideal husband/wife operation. Just be sure that you and your spouse agree and get your signals straight. Failure to do this has often been a source of petty annoyance for me with some of my clients. Problems arise because of poor communication.

Get What You Want

This reminds me of a thought I will pass on to you. You will get more of what you want in your house, with fewer hassles, when you act as your own general contractor. Most contractors are very staid.

They like to do things the same old way. This often causes problems for the buyer who wants something done a little differently. With you as the general contractor, you get things done as differently as you like.

One very simple and not too costly step you can take to save time and ensure better communication is to put a phone on the building site. Long distance calls are easily controlled. In my many years of contracting I've never had the job phone abused by a sub. It has saved me countless trips to the job site, too. Just be sure you or a sub takes the phone home each night.

Years ago I built a vacation home in the mountains about 140 miles from our permanent residence. I made only three trips to supervise and check on the progress the entire time the house was being built, from the staking of the lot to the final inside trim. I don't recommend this. I only mention it to show that you don't have to be there every minute or every day. My mountain house was built by phone. This was far cheaper and much less tiring than driving 140 miles each way. The subs did a beautiful job. They wouldn't have done any better if I had been there watching them.

Work by Phone

Phone calls to subs are the key to this business. Make those calls before they leave for work, or in the evening when you are not working. On-site inspections can be made by you before or after work or on your lunch hour. Daily on-site inspections usually aren't necessary. If you believe they are, and you can't do it, your spouse or a friend can. You don't have to be there every minute. There isn't a general contractor alive who is. For me that would mean that I could only build one house at a time and when I was building, I usually had five or six under construction at once.

You don't have to watch masons lay every brick or your carpenters pound every nail. You allow them to do their jobs. Mistakes may be made from time to time. Chances are they would have been made if you were there. There isn't a mistake in the whole process that can't be rectified. It should be your individual decision on how much time you want to be at the site.

You Can Get a Loan

Unless you are paying cash for your house, you will need a loan. Almost all loans for houses are made by banks, mortgage companies, and credit unions.

For all practical purposes, getting your loan is the most important step, and the most crucial in building. No money, no house. You must determine very early in your decision-making process whether you will be able to get a loan.

The intricacies of financing are explained in Chapter 3, where we discuss how a loan works and how much you can borrow. I will be very honest and say that most lenders will be somewhat reluctant to lend to an individual who plans to be his or her own general contractor. This reluctance is legitimate. Lenders want to be sure the house will be built properly, and most important, finished. They don't want to step in and finish a house. They are lenders, not builders, and that's the way they would like to keep it. It will be a challenging sales job convincing them you can be your own general contractor, but with the knowledge you will receive from this book, you can do it.

Plan your visit carefully. You want to appear to be a person who has thought through all steps in the construction of the house, who understands the problems and has solutions to them. You know the lender won't be enthusiastic about your decision to be your own contractor, so you must do all in your power to present this approach as an asset, not a liability, and your projected savings can make it an asset.

The lender will wonder whether you are an organized person who can handle the details of home construction. You will be convincing if you have all of your homework completed when you reach the bank. "Make it look as though the only step left is to get the money," suggested one lender.

Information You Need

Your presentation should include a package of material, including the following:

1. Proof of your current income. Include joint incomes from all sources.
2. Good credit references.
3. A financial statement of what you own and what you owe.
4. Evidence that you have sufficient knowledge about being your own general contractor.

Another Lender's Views

Lenders, I've learned, don't differ too much across the country. They tend to be wary when you mention being your own general contractor. They're willing to be cooperative if they think you have a good chance of finishing the job, but they don't want to be stuck with a half-completed house.

A New England lender quickly rattled off the three points we've already listed as requirements, covering credit references, a financial statement, and proof of income.

I asked what would make him more warmly disposed toward the would-be general contractor. He listed these points:

- Own your lot, free and clear.
- Show proof of your ability to handle the construction, as described in this book.
- Have as much planning completed as possible to assure the lender that this is not an idle dream, but a project that only lacks the needed financing. Such planning would include a set of house plans, tentative arrangements with suppliers (including estimates of the costs of the materials), and all information needed on such subjects as availability of utilities, any zoning restrictions, and regulations of the building inspectors.

Numbers 1, 2, and 3 are prerequisites to buying or building under any circumstances. Number 4 is what this book is all about. The lender will check out your knowledge of the topics covered in all of the chapters, and will easily recognize your ability to estimate, find a lot, and select plans, and your knowledge of construction loans, permanent mortgages, subcontractors, suppliers, and the different phases of construction. Don't worry — all of these items are explained in this book.

Don't Be Discouraged

After familiarizing yourself with the material in this book, there should be no reason for you not to get a loan. However, you should keep in mind that you may get turned down by one or two lenders before you get an approval. Don't be discouraged. Sometimes it is nothing more than personal chemistry, or other trivial things, and has nothing to do with your ability.

It costs nothing to go through the initial stages of discussion with a lender, and you may get a commitment for a loan. In building my first house, I was turned down by four lending institutions before I found two who would permit me to be my own general contractor. If I hadn't been persistent, who knows where I would be right now. I was turned down because I didn't know everything you'll know when you finish reading this book. I didn't show sufficient knowledge until after the third rejection. Each time I was turned down, I would do a little more homework and had more confidence in myself. I didn't have a book — you do. You can do it.

An Alternate, The Manager's Contract

When you act as your own general contractor you must do a good selling job to the lender. And, after that, let's suppose the worst happens. You can't get a loan with yourself as the general contractor. Or, let's suppose you simply don't feel able to deal with subs.

Here are three other ways you can build, each using a general contractor in a position of increasing responsibility and cost to you:

1. A manager's contract.
2. Cost plus a percentage or a fixed fee contract.
3. Contract bid.

These are listed in the order that they usually increase costs to you. Since each one increases the contractor's responsibility, it will increase the cost of having that contractor. The only things that will not fluctuate are profit and overhead. The cost of the house, land, and other fixed expenses should remain the same.

For all practical purposes, options 2 and 3 are readily available contracts and are used by most builders with varying degrees of legalese and varying degrees of slant favoring one or the other parties. (Samples of contracts for all three options are shown in the Appendix.)

The least expensive way to go — and one by which you can still be considered the general contractor, be accepted by your lender, and possibly feel more comfortable dealing with subs — is using the manager's contract, option 1.

Under this arrangement, you have a licensed general contractor who has one responsibility, to act as your manager with the subs. This contractor, in return for about one-third to two-thirds of his normal fee or profit and overhead, will assist in finding the subs (although you can still find your own), will schedule subs, check the quality of their work, approve the quality of materials, and order materials, when needed, in your name. You will be responsible for securing suppliers, permits, loans, paying all bills, including subs, and inspections for quality and approval. You will be responsible for the final job and its overall acceptance.

Some of the advantages of this contract are that the general contractor will:

- Do a cost estimate.
- Help find subs.
- Schedule your time better.
- Help cut red tape and get the necessary permits.
- Arrange for temporary services.

- Schedule subs, from survey to landscaping.
- Develop materials lists.
- Assist in reviewing bills when requested.

A slight change from this procedure would be to hire a designer-builder who would advise you on plans, make any necessary modifications to those plans, review the specifications for such things as adequate insulation or the possible use of native materials, and inspect the process of construction.

In either case, the manager's contract shown in the Appendix could be modified to include more or fewer of the responsibilities of the general contractor. The cost, of course, will vary with the number of responsibilities listed.

If you decide to use one of the contracts in the Appendix, go over it with your attorney to check it for applicability and to make certain it conforms with local laws.

Chapter 2

Where to Start

YOU'VE MADE YOUR decision to build. Wonderful. You'll do well, so don't worry. What comes next? Well, where do you want to live, based on your needs and desires and finances?

Remember, in making this decision, that you can use this house as a stepping stone. Once you've built your own home, you can move up to a bigger or better house or location and possibly at no — and I mean no — extra cost. Each time you move you are adding to your profit as a builder, and thus are increasing your own *equity.*

Land

If you have lived in a city or town for a while, you should know where you want to build. If you haven't looked around, or if you are new to an area, I suggest using a local real estate broker. These brokers know what each area offers and what lots cost in different locations. If acreage is what you are after, here, too, a real estate broker is most helpful. Don't forget, your broker is one of those helpers I mentioned. A broker can help you locate the property you want, and can help with all the details necessary to assure you that it is a suitable building site. The broker should be able to show you a map of the lot, and point out the boundaries to you during a walk around the lot.

If you don't already own land, first choose the area where you want to live.

Next, look for a lot or a site with acreage in that area, one that is suitable to build on.

Sloping Lots

If you want a basement and you live in an area of the country as I do where the soil doesn't drain well, you should avoid flat lots. A sloping lot will provide drainage by means of footing drains. If you

Do You Need a Basement?

Do you want — or need — a basement?
Here are the arguments for and against constructing a basement.

For:

- It's relatively inexpensive, compared with aboveground space, and particularly if you must dig down six or more feet anyway for placing footings and foundation walls, as you must in some parts of the country.
- A basement gets your heating system, and sometimes the fuel supply, out of the way.
- It's a good storage space — but may be too damp for paper, metal, and clothing.
- A basement may provide you with garage space.
- It's an excellent space for an out-of-the-way family room.

Against:

- If the water table is high, flooding may be an annual headache.
- A basement may be too damp for many uses, or require a dehumidifier.
- It's an ideal spot for storing things better taken to the dump.
- Building a basement can be expensive, particularly if the lot has an underground ledge or the area is sometimes flooded.

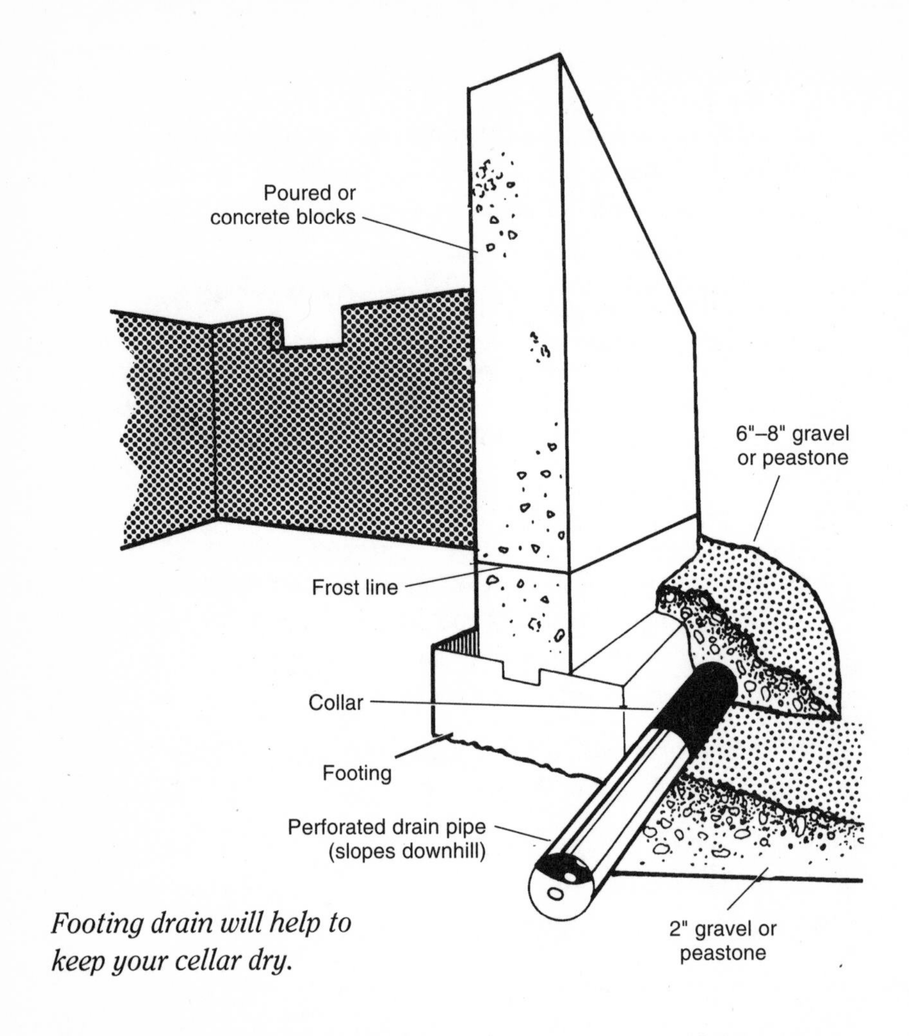

Footing drain will help to keep your cellar dry.

want a basement, sloping lots are best in any part of the country. They assure drainage no matter what type soil, and they allow for one side or more of the basement to have good window areas for light and dryness. Also, in some areas the open side can be *frame construction*, which is a little less expensive than foundation brick or concrete block.

If you don't want a basement, try to find a relatively flat lot so that you won't have an excessive amount of *crawl space* (or fill, if you decide on a *slab foundation)*. Crawl space is cheaper than a basement, and if the lot slopes, even excessively, only on one end or one corner, the cost of accommodating a foundation to that lot should not be increased too much. Notice I say "accommodating a foundation to that lot." So many people put the horse before the cart. I can't say strongly enough: find the land first and then the plans to fit.

In colder regions, south-sloping lots are becoming increasingly popular. They're ideal for solar homes, and offer protection from the chill of north winds.

Trees Are Valuable

If you like trees, try to find a wooded lot within your budget. Barren lots are cheaper to build on, but more costly to landscape — almost a direct cost wipe-out. A wooded lot generally costs more to buy, more to build on, and less to landscape. It's your choice, but if you are concerned about top dollar resale in the near future, it's worth putting a few thousand dollars extra into a wooded lot.

If the area in which you are looking has no development activity near it, I most strongly recommend *test borings* of the soil, before you purchase the lot, to determine its *load-bearing capabilities.* It will also show whether there is ledge on the site that might require blasting. This test is not expensive and one the seller should be willing to pay for. I have spent hundreds of extra dollars on *footings* due to poor soil. In some parts of the country, I'm sure that could have been thousands. Be sure that this test is included as a contingency in any contract to buy. It can also be included as a refund provision in the contract in the event that non-load-bearing soil is discovered after purchase.

If sewer and water facilities are not available on the lot, the contract should have a contingency as well as a refund provision regarding septic tank and well use. In some states, this is a law. Land must be suitable to accommodate a septic system. The suitability is determined by local codes, but generally it is based on how quickly the soil drains. A soil test determines suitability.

Specialists in test boring for load-bearing are listed in the Yellow Pages under "Engineers, Consulting" or "Engineers, Foundation." Most county health departments make soil tests for septic systems. This is done for free, or at a very low cost. These departments also provide information about wells.

Water and Sewage

Both water and sewage must be considered as you select your building lot.

If water and sewage are both provided by your city or town, you need only find out how to tap into them, and what are the costs involved.

If either one is provided (and quite often it is only water), your problem is slight. You may need some form of a permit, plus a percolation test for a septic system, to determine whether the soil will absorb the discharge of the system.

If you must provide both water and sewage, some study is needed.

A septic system consists of a sewage pipe running from your home to the septic tank, which is usually a large concrete box. Sewage enters this box and is broken down by bacterial action. A liquid then flows from the tank into a system of perforated pipes that permits the liquid to seep out underground in an area called a leach field. The size of the field required depends on the

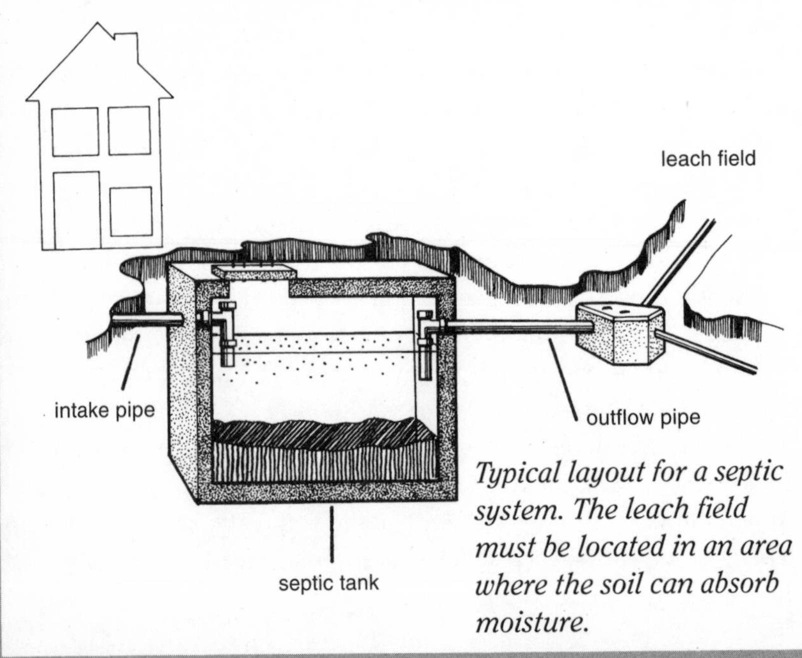

Typical layout for a septic system. The leach field must be located in an area where the soil can absorb moisture.

amount of sewage going into the tank, and the ability of the soil to absorb moisture. Sand is best; clay creates a problem.

Regulations in many regions require that a well be a certain distance, say a minimum of 150 feet, from your leach field, or from the field of any neighbor. Thus it's best to have a large lot if you must provide both water and a sewage system.

Digging a well can be expensive. You can get a dowser, complete with forked branch, to "locate" water for you, or you can do what most well diggers do, and that is to locate the well where it's most convenient, and start digging.

Most well diggers charge by the foot of depth they bore, and usually can estimate about how far they will have to dig. Some diggers will offer you a contract price for the well.

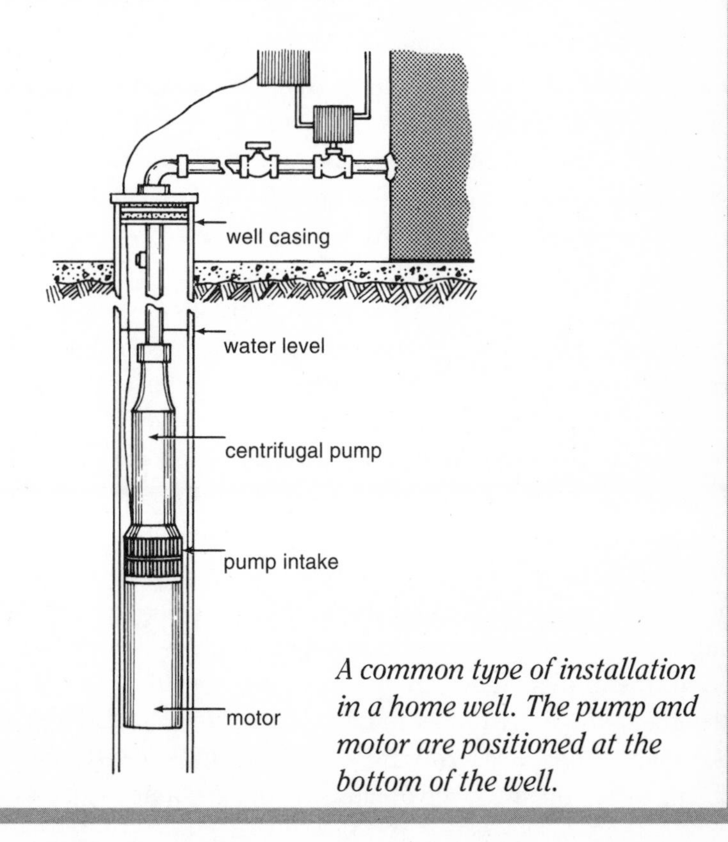

A common type of installation in a home well. The pump and motor are positioned at the bottom of the well.

Let the Brokers Do the Work

Put the burden of getting these tests on your real estate brokers. Let them do the legwork and checking. Just make sure it's in the contract you're offered. Let them earn their commissions. Let them hand you a nice, clean finished deal — a lot ready for you to build your dream house on.

If you deal with a realtor — a member of the National Association of Realtors — you have a recourse if there is any problem, and that is the association. In almost all towns and cities the local Board of Realtors is so image-conscious and worried about a member ruining that image that it will protect you, the buyer. Simply call the board if you believe a realtor isn't doing his job, or isn't representing your best interests.

Zoning

Consider *zoning* when you are choosing a lot or acreage. Be aware of what could come into your future neighborhood — stores, offices, trailer parks, or industry. Look at what is already there. Such things as dumps, railroads, and industrial buildings make the lot worth less as a site for a home. One could write a book on zoning — and many have. A quick consultation with your real estate broker and your attorney should resolve any fears you have.

If water and sewer are provided, be sure your real estate broker (or you, or your attorney, if no broker is used) checks out all costs to get the water and sewer to your property line if it is not already there (it may be across the street), and any and all tap-in fees or privilege fees charged by the municipality or association providing the services. Do the same for gas, electrical, and phone services.

Checking out wells, septic system, load-bearing capabilities, water and sewer fees, and zoning may take a few weeks. If you like the lot and don't want to risk losing it, you can make an offer with the understanding that your binder or deposit will be refunded if the conditions mentioned above are not in your favor. Don't put more than 10 percent down as a binder or deposit, and if using a broker, be sure the money is held in escrow. Above all, consult your attorney before closing a deal on any piece of land. I could say that 100 times, and it

Checklist for Buying a Lot

Here are some of the many things you will want to check before buying a lot:

Neighborhood
- Will your house fit in with others on street? ________
- Quality of schools? ________
- Attractive street? ________

Transportation
- Public transportation available? ________

Pollution (air, water, noise, such as airport) ________
Zoning restrictions ________
Sewage, water available ________
Soil tests made ________
Direction of slope, if any ________
Drainage (observe after heavy rain, if possible) ________
Test borings ________
Size of lot ________
Trees (valuable to have, but sun needed for garden) ________
Title Clear ________
Cost (A counter offer of 10–20 percent below asking price is often accepted.) ________

wouldn't be enough. Have him or her check the title to the lot, to make sure it is clear. Check out all restrictions on the size and type of dwelling you can build, and what a neighbor may build. They may or may not be covered in zoning. Different parts of the country use different means to protect an area from visual blight. Again, your broker and your attorney can help you on this.

How Much Should You Spend?

Land costs probably vary more than any other item in the construction business. Three things determine the selling price of land. They are location, location, and location. I'm sure you've heard that before, but it's true. However, the more you spend on land, the less you'll have left over for the house.

You will need to determine building costs in your area, plan a total budget, and then shop for land based on what you have left after subtracting building costs. Building costs can be quickly determined in any area of the country by using my "windshield estimating" method described in the estimating chapter. If your total budget (cash + mortgage) is, for example, $100,000 and you find a piece of property that sells for $40,000, and local building costs are $50.00 per square foot (with you as the general contractor), that means you could only build a 1200 square foot home. $100,000 – $40,000 = $60,000 ÷ $50 (per square foot) = 1200 square feet. Not a very big house for such an expensive piece of property.

It's a juggling process between land cost and house size, but ideally, the land should be approximately 25 percent of the fair market value of the house and land together.

Buying land is a very subjective process. One tip I'll pass on, however, is to always keep resale value in mind. I know that may not seem important to you, especially since selling is the furthest thing from your mind right now, but some day it won't be. It is always on the mind of your mortgage lender.

The House Plans

There are two ways of getting house plans. One is to find a plan suitable for you and your building site by looking through magazines or books of plans. The other is to have plans drawn for you by an architect or drafter/designer.

The least expensive way to obtain plans is the first one. There are thousands of magazines and books of plans. Quite often you'll find that one of the plans needs only minor modifications to make it suitable for you.

You either should stick with the plans the way they are or order the minimum number sold (sometimes this is only one) and get advice from a drafter/designer or an architect on any changes, no matter how minor. This will be much less expensive than having the entire plan drawn for you. A drafter can even make major changes in a set of plans, and can advise you on the practicality of changes you suggest, and the additional cost, if any, of those changes.

Make Changes Early

Try to make all changes in the plans before you start construction. You need not have the plans redrawn for minor changes such as moving, adding, or deleting a window or door. But you should if you move walls, or change roof lines and roof pitches.

If the drafter/designer redraws the plans, it will cost about twenty to thirty cents per square foot for a finished product, including six to eight sets of the plans. If he or she designed your house from scratch, you could expect to pay from thirty to fifty cents per square foot. In that case you usually would get more precisely what you wanted.

You could hire an architect to do this work. This would cost considerably more. Either one can do the job and either one should give you a quote over the phone of approximate costs. Look for names in

Cutting Construction Costs

Your first chance to save a lot of money is now, when you are selecting the plans for your home.

- Do you need that many square feet? Every reduction you can make in size will cut the costs.
- One story or two? It's cheaper to build the same number of square feet of housing in a two-story house.
- What is its shape? A simple rectangle is the cheapest to build. Additional corners add to the costs.
- What style roof? The least expensive is the common gable roof.
- How's your timing? If you're building when there's a lull in construction (i.e., winter), chances are you'll get your money's worth — and maybe a bit more.

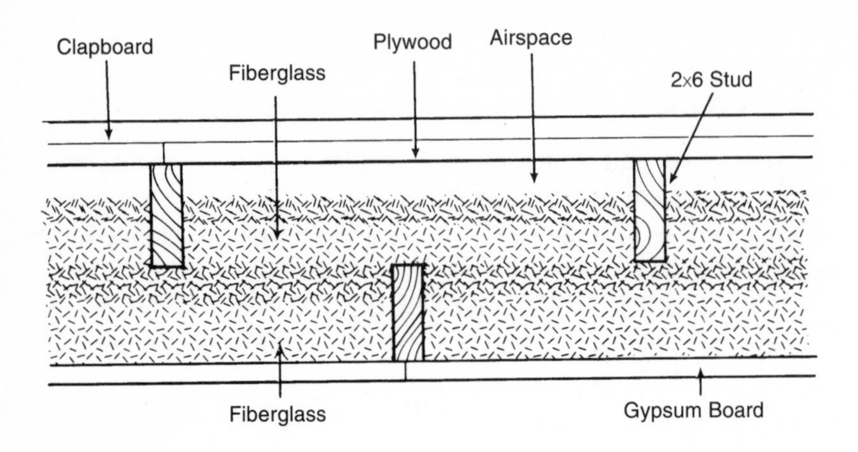

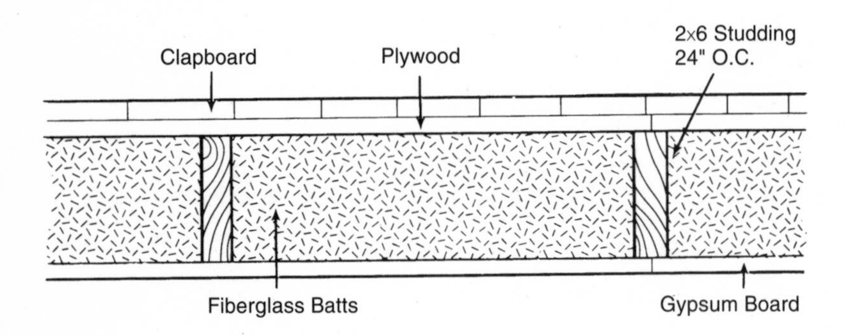

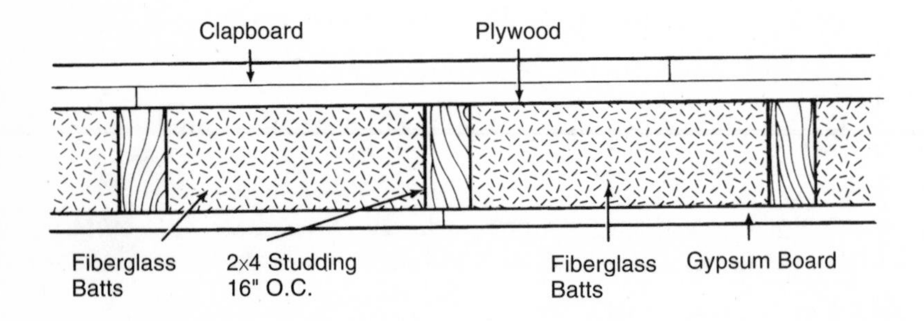

Three ways (and there are many more) of building outer walls, and their R factors. At top, this has an R factor of about 33, with the 2×6 studs not touching both the inner and outer layers. Center, with 2x6 studding, is about R-22. Lower is about R-13, and shows the conventional method used until a few years ago.

Insulation

You're lucky. While those of us who own homes built before 1970 wish we had a chance to increase the insulation in them, you are starting fresh, and can plan a home that is energy-efficient.

Consider this as you select your plan, and as you talk over your plans with your subs, particularly your carpenter and your insulation sub. There are decisions you can make that will result in thousands of dollars of savings as you live in your home over the years. Some decisions will require changes in your house plans and specifications.

If you have a full basement, should you insulate it? Do your windows provide two layers of glass? How will your outer walls be built? (Three possibilities are shown here.) How much insulation will you have over your head? Will your heating ducts and hot-water pipes be insulated? Is your chimney inside the house, where any heat it collects will be returned to the house, or on an outer wall, and thus providing a high heat loss? Does your fireplace provide a source of outside air, so that you're not wasting heated air, and does it have a tight damper, so hot air isn't flowing up it when the fireplace isn't in use?

You'll find utility companies discuss R factor. This refers to the ability of a material to resist the escape of heat. A high R factor means that you are not losing as much heat. There are standards of R factors for walls and roofs that vary geographically.

the Yellow Pages or get recommendations from friends. Shop for price. Before you select an individual or firm, ask to see work done, and be sure to ask for a few references. Don't be timid and most of all, don't be intimidated. After all, you will be the employer.

On-the-job changes are expensive, so make your decisions on paper and live with them or expect high cost overruns. I usually rough sketch my houses on quarter-inch graph paper, and make most of my changes in this stage. If you plan to start from scratch with your plans, you can do this too. A twelve-by-twelve-foot room will be drawn three squares by three squares. This doesn't allow for wall thickness, and I don't worry about it in the early planning stages unless there's a tight squeeze with a room or a stairway and I need to know then that it's going to fit. If this planning makes you feel uncomfortable, don't do it. Buy your plans or have them drawn for you from the sketching stage up. I only do it because I enjoy it. It really is a job for one of your subs, your architect or drafter/designer.

Stay closely involved with this phase, even if you don't sketch the plan. Pay strict attention to what is in the floor plans. Picture yourself walking through the house from room to room. If in doubt as to whether a room is large enough, find a room of about the same size, and compare.

This is a good time to mention the only two tools you'll need as a general contractor: two steel tape measures, one twenty-five feet long, and the other 100 feet. You may not use them many times, but they will help you eliminate guesswork in the planning stages.

Studying Plans

Figuring out the plans is easy, although it may look difficult at first. You'll be looking first at room sizes, room placement, traffic flow, kitchen work flow, closet space, number of baths, and over-all size. Those are the major functions of design. Without too much effort you can move doors and windows, add them, or delete them — on paper.

Unless you have an unlimited budget, and very few of us do, a house is a series of compromises. We can't afford all of those beautiful things we see in idea magazines. A house is going to cost so much per square foot using the average number of windows and doors and average appliances. If you want a $4,000 Jacuzzi bath and can't afford to spend $4,000 more, compromise on something else — the size of the house, the garage, or a paved driveway.

Don't make any of your decisions a "matter of life and death." You may be surprised how unimportant that "big decision"was when you are finished. Agonizing over decisions only leads to friction —

within yourself and with others. Make all of your decisions conscientiously but quickly, and then move on.

Be sure plans include specifications (specs). Specifications are lists of materials that are going into your house. Most plans come with a set of specs. Your drafter or architect will include a set with the drawings.

There will be a "for input" in the specs for you. (Refer to the "Description of Materials" form in the Appendix (page 107) and you will see that there are places indicating decorative items, such as wallpaper, carpet, paint, stain, and door hardware.) Obviously, you will choose these. Since specs come early in the game, most people haven't selected these decorative items at that point. Therefore, monetary allowances are used with the term "or equal." This means that the actual item finally selected, such as a kitchen faucet, will be of an approximately equal dollar amount and of similar quality to the one in the specs.

Parts of Plans

Be sure your plans include the following pages, all of which are shown here or in the Appendix.

- **A plot plan.** If your plans are drawn locally, the designer can prepare this plan for you. Otherwise you or a surveyor will have to do it. It is merely a plat or map of your lot with the position of the house drawn or blocked in. Its purpose is to insure that a given house plan will fit on a given lot. If it doesn't fit on paper it surely won't fit in reality.

  First, the lot is drawn, then all the setbacks required by zoning and/or restrictions are sketched in, then the location of the house is indicated. The exact location may be slightly different when the house is physically staked out, but will be close to the indicated plat position — especially if it is a tight fit. If you have a lot consisting of several acres, the actual position of the house from that indicated on the plat could be changed by many feet.

  The plat or plot plan won't be drawn first if you have a set of plans but no lot. It's better to find a lot first, then plans to fit. That is not an absolute rule. Some people love a set of plans or they want a particular house they have seen in a magazine. They would rather search for a lot to fit the plans. It can be done.

- **Spec sheets.** In these lists of materials for your house, spell out as much detail as possible. Using an existing set (such as those found in the Appendix on pages 107–118) is the easiest way to start. Make any necessary modifications.
- **Foundation plan.** This shows the overall dimensions of the house and the locations of all load-bearing requirements, such as *piers*, steel, *reinforcing rods*, vents, basement slabs, and if a basement house, all walls, windows, doors, and plumbing of that basement. (In the Appendix you will find a foundation plan.)
- **Floor plans.** You must have a plan for each floor. It will show outside dimensions, plus locations of windows and doors, plumbing fixtures, and large appliances. You may add the electrical drawings as well. These drawings show the locations of receptacles and switches. It is not necessary to have these shown. After a house is framed, you can go through it with an electrician and position receptacles and switches. However, for the purpose of getting quotes or bids from subs, it is best to have the electrical requirements included either in the floor plans or on a separate sheet. The same is true for heating and cooling (mechanical) requirements. It is best to have these drawn out in advance, for bid purposes and to locate the furnace. I use heat pumps exclusively in my houses so I don't have to worry about furnace placement. With wood, coal, gas, or oil planning the *flue* location is necessary, especially in a house without a basement and/or in a two-story house.
- **Detail sheet.** This is a sheet showing cabinet details, cross sections of the inside of the house if there is a section that may not be perfectly clear from the floor plan, wall sections to show all the materials that make up a wall, such as brick, sheathing, studs, insulation, inside wallboard or paneling. A

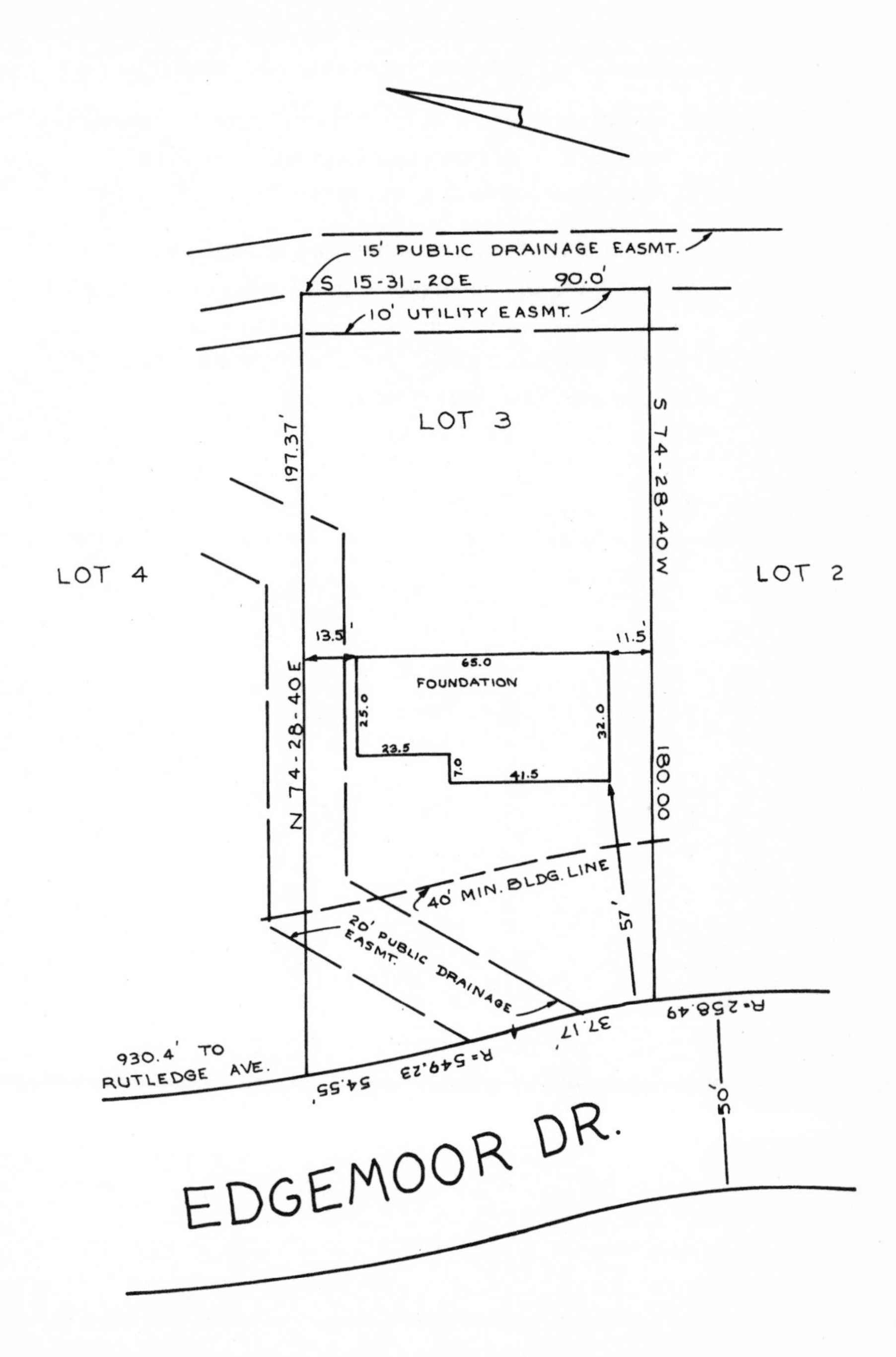
15' PUBLIC DRAINAGE EASMT.
S 15-31-20E 90.0'
10' UTILITY EASMT.
LOT 3
197.37'
S 74-28-40W
LOT 4
LOT 2
13.5'
11.5'
65.0
FOUNDATION
25.0
32.0
23.5
7.0
41.5
N 74-28-40E
180.00
40' MIN. BLDG. LINE
57'
20' PUBLIC DRAINAGE EASMT.
930.4' TO RUTLEDGE AVE.
R=549.23 54.55'
37.17'
R=258.49
50'
EDGEMOOR DR.

Plot plan.

detailed section of a foundation wall is sketched to show proper construction technique, drainage, and waterproofing.

- **Outside elevations.** These are the sheets (usually two) that show an outside view of how the house will look when finished. Usually all four sides are shown.

You'll need about six sets of plans, one for yourself, one for your lender, and one for each of the major subs. If you can get sets of plans that are inexpensive, it's good to have as many as possible. If you're going out for bids, for example, for work by a sub, each of the bidders should have a copy of the plans to study.

Chapter 3

Financing

IF YOU FOLLOWED the instructions in Chapter 1, by now you have convinced a bank, mortgage company, or credit union to give you a loan. That loan will be divided into two phases:

1. The construction loan.
2. The permanent mortgage loan, to be paid off in twenty to thirty years.

Usually any difficulties in obtaining a loan involve the first phase, the construction loan, not the permanent loan given when the house is finished.

The Construction Loan

If you were able to secure a $60,000 mortgage for your new house and the lender were to give you a check for that amount before you started, both you and the lender would be in trouble. The lender, because they just loaned $60,000 on a house that doesn't yet exist; you, because you'd have to pay interest and principal on $60,000 before you even broke ground for that new house.

Considering interest expense, that could mean you would have to pay a large amount of interest each month on a nonexistent house, along with your regular monthly fixed expenses for items such as food and rent.

Enter the construction phase of mortgage loans —the construction loan. When you are approved for a mortgage for a house yet to be built, the lender will set up your loan first on a short-term basis, called the construction loan. The money is disbursed to you as the house progresses.

Only Interest Paid

You only pay interest, no principal, on the total amount lent to you for any given month during construction. For example, if the house is 20 percent complete after the first month, the lender, after a physical inspection of the progress of construction, will advance 20 percent of your total to you to pay your bills. If your total mortgage is, say, $60,000, then you will receive $12,000.

The house is now 20 percent complete, you are one month along in progress, and you have paid no interest yet. And you won't have to until next month, when you will then pay the interest, say 10 percent for thirty days, on the $12,000 you received. That payment will be $100.

This is extremely important. Many people are afraid they will face two house payments at the same time. Not so. You use dollars that were advanced for construction to pay the monthly interest. You make no repayment of principal. You do not use your money. Construction loan interest is a cost of construction of any new home, whether you build it or I build it for you. It's not even part of the builder's overhead. It's a cost like lumber and nails. It is the second item on the estimate sheet in Chapter 4. Treat it as such and plan for it, along with other financing charges, and it won't bother you. Think of all the money you are saving by being your own general contractor.

There is a considerable time period during construction in which the house progresses toward completion, but you haven't received bills for materials. It will take you approximately six months to complete the house, and because of the billing time lag from suppliers and loan interest being one month or more in arrears, in that six months you will have had an average of only 40 percent of the total money available disbursed to you. This is an approximate figure, obviously, as is the time frame needed to complete your house. I'm merely pointing out that you won't pay interest on the full amount of your mortgage until the house is complete. At that time the lender will complete the disbursement of the construction money and convert the loan to a permanent mortgage for twenty to thirty years. Then you will be making your full payments of principal and interest. But then you have your dream house. Move in. If that sounds simple, it is only because it is.

INSPECTION REPORT AND DISBURSEMENT SCHEDULE

Borrower ______________________________

Contractor ______________________________

Job Address ______________________________

City ______________________________

DATE OF INSPECTION							
1. Excavation	2						
2. Foundation & slab work	8						
3. Floor	4						
4. Wall framing	6						
5. Roof framing & sheeting	6						
6. Wall Sheathing	1						
7. Roofing	2						
8. Windows set	10						
9. Exterior doors	2						
10. Siding/brick	7						
11. Exterior trim	1						
12. Well/water tap	2						
13. Septic sewer tap	2						
14. Plumbing roughed	5						
15. Wiring roughed	2						
16. Heating roughed	2						
17. Heat plant	2						
18. Insulation	2						
19. Chimney/flue	2						
20. Drywall	6						
21. Bath tile	1						
22. Plumbing fixtures	4						
23. Interior trim	3						
24. Cabinets	3						
25. Exterior painting	2						
26. Interior painting	2						
27. Appliances	2						
28. Light fixtures	1						
29. Carpet/floor finish	4						
30. Drives & Walks	3						
31. Landscaping	1						
Total	100%						

Inspector ______________________________

Comments:

A Choice of Mortgages

Here is a short description of some of the mortgages currently available.

Conventional mortgage. My favorite because it involves the least amount of red tape, and usually is fastest to get. You may have to have a larger down payment, depending on the laws in your state, and the requirements of the lender.

Federal Housing Administration Mortgage. FHA insures the loan. The down payments required are small, the repayment period is up to thirty-five years, but it may take time for you to get your loan approved and for FHA to check out your plans. Your lender will have specific details.

Veterans Administration Mortgage. You have to qualify as a veteran to get the VA to guarantee your loan. No down payment is required in some cases. Approval is sometimes slow. Ask your lender for specific details.

Farmers Home Administration Mortgage. There are both guaranteed mortgages and loans made directly from FmHA funds. The interest rate you pay depends on your family income and the size of your family. This can be an excellent way to borrow for a low-income family. Check this one by telephoning

Two Homes?

By the way, even in a sluggish market, houses sell sooner or later. If you already have a home and you fear it won't sell before your new one is ready, either don't worry about it or sell it first and live in an apartment while building your new home.

The other financing charges will vary with lenders and attorneys, but can be estimated in advance. They will total about 2 to 3 percent of the mortgage amount. They may have to be paid when you close your construction loan, or they may be deducted from the $60,000.

I prefer the latter method. It doesn't tie up my cash. They also are a construction cost.

the FmHA, listed in your telephone directory under U.S. Government, Agriculture Department.

There are a number of other ways of financing homes that have developed since the late 1970s when interest rates were high. Depending on the state of the economy when you are seeking a mortgage, there may well be additional options you will want to investigate.

The most simple is the *mortgage financed by the seller.* Used more often for home sales than for lot sales, this mortgage has terms negotiated by the two parties involved.

The *graduated payment mortgage* is tailored for young people with expectations of higher incomes. Monthly payments are small at first, but gradually increase. Figure out total payments under this mortgage, and compare them with those of a conventional mortgage before being swayed by those smaller early monthly payments.

The *variable-rate mortgage* can be written so that monthly payments or the total number of those payments can be changed during the life of the mortgage as interest rates move up or down.

Cash Needed

You will need some extra cash. Call it interim or short-term financing. Some expenses will occur before your lender makes the first disbursement of construction money and you will need a few thousand dollars to bridge that gap for such things as paying subs, fees, and permits. If you have cash available, fine. If not, you may want to borrow from a commercial bank or take a second mortgage on your existing home, or a *bridge loan.*

Excluding the cost of the land and the 2–3 percent we discussed, you shouldn't need more than an additional 5 percent of the cost of the house for interim operating expenses. This money is needed only until the house is finished. When the lender disburses all the money in the form of the permanent loan, the need for your operating cash will disappear and you can put it back where you got it.

If it is borrowed, include the interest paid in your cost estimate. It, too, is a cost of construction that other builders and I have included. It may seem that I put the cart before the horse by discussing interest charges first, but many worry about them and I want to allay that worry.

FHA and VA loans are government-backed and they deal *only* with permanent financing and not construction financing. Your local lender will coordinate the construction loan.

Usually the lenders will require that your land be paid for in full and that you have a *clear title* to it before they will lend you money to build on it. This is to protect them. Quite simply put, if you don't have clear title to the land, they won't have clear title to their *note* to you for construction money.

There is a whole profession handling those matters, so we won't go further into them than that. Your attorney can explain the legalese far better. It is not necessary for you to become versed in law to be your own general contractor.

Lot Subordination

There is a way to avoid paying for the land before construction. It is called *lot subordination.* It means the owner will take a note in lieu of payment of all or part of the asking price, and legally subordinate his interest in that land to your lender. Consult your attorney for details. Some lenders won't allow this or any other form of borrowing for the land or down payment. Usually the land is the down payment as its value is close to the normal dollar amount put down on a house. Lenders want that money to be your *equity.* They may have you sign a statement that you aren't borrowing your down payment.

But in your case, you are saving an amount equal to your down payment or your land by being your own builder. The builder's profit and overhead you are saving is a legitimate figure. You will earn it and it will be an actual dollar amount when the house is finished.

Lenders will have different opinions of many of your particular financing needs. You will need to shop around and spend some time talking to them. They will help you tremendously in putting your financial package together. In this aspect, they are like a subcontractor. Each of them will also explain their policies on construction draws, inspections, construction loan interest payments, points, credit reports, qualifications, and interest rates.

The Permanent Mortgage

Once you've secured funds for construction, there are several things to keep in mind as you seek a permanant mortgage.

Qualifying for a Loan

As to qualifications or "qualifying for a loan," the requirements change daily and are different for each lender. A rule of thumb that is often bent is that your annual mortgage payments (only on your permanent mortgage) should not exceed 2 to 2½ times your gross monthly income from all sources and from both spouses. Local lenders will have various rules of thumb, but most will be about like this. Also, the more money you put down (increasing your equity), the more willing a lender will be to lend you money, and often at a lower interest rate.

Don't forget, the thousands of dollars you are saving acting as general contractor are a legitimate value, and are to be, under all circumstances, considered as cash equity.

Keep this in mind when talking to lenders: due to more and more people being squeezed out of the housing market each day by high prices and interest rates, lenders are making fewer loans. Yet they must make loans to stay in business, since it is a major source of their income. Perhaps you might have been squeezed out too, in the past, but now, by being your own general contractor, you can handle a loan. The difference could be that money you'll save by going this route.

How Big a Mortgage?

How much house you need and therefore how much you need to borrow is up to you. Chapter 4 offers a fast, easy, and fairly accurate method of estimating the costs. Stretching your budget for more house is not all bad, because:

- You get a house large enough for a growing family.
- You are making a good investment. Houses appreciate in value while you get an excellent tax write-off on the tax-deductible interest.
- Your income will probably catch up in a few years. Check into graduated mortgage payments or variable interest rates.
- You can always bail out by selling the completed house and recoup all costs plus a profit. Then you're ready, of course, to reinvest in another new home.

Financing may be the most difficult phase of construction as it will be the one in which you may feel the least adequate. You will find, however, that after talking to one or two lenders, you'll be more relaxed and better versed in their jargon. Of course, you may get lucky and establish a good relationship with the first lender you approach.

STANDARD MORTGAGE APPLICATION FORM

Uniform Residential Loan Application

This application is designed to be completed by the applicant(s) with the lender's assistance. Applicants should complete this form as "Borrower" or "Co-Borrower", as applicable. Co-Borrower information must also be provided (and the appropriate box checked) when ☐ the income or assets of a person other than the "Borrower" (including the Borrower's spouse) will be used as a basis for loan qualification or ☐ the income or assets of the Borrower's spouse will not be used as a basis for loan qualification, but his or her liabilities must be considered because the Borrower resides in a community property state, the security property is located in a community property state, or the Borrower is relying on other property located in a community property state as a basis for repayment of the loan.

I. TYPE OF MORTGAGE AND TERMS OF LOAN

Mortgage Applied for: ☐ V.A. ☐ Conventional ☐ Other: ☐ FHA ☐ FmHA | Agency Case Number | Lender Case Number

Amount $ | Interest Rate % | No. of Months | Amortization Type: ☐ Fixed Rate ☐ GPM ☐ Other (explain): ☐ ARM (type):

II. PROPERTY INFORMATION AND PURPOSE OF LOAN

Subject Property Address (street, city, state, ZIP) | No. of Units

Legal Description of Subject Property (attach description if necessary) | Year Built

Purpose of Loan ☐ Purchase ☐ Refinance ☐ Construction ☐ Construction-Permanent ☐ Other (explain): | Property will be: ☐ Primary Residence ☐ Secondary Residence ☐ Investment

Complete this line if construction or construction-permanent loan.

Year Lot Acquired | Original Cost $ | Amount Existing Liens $ | (a) Present Value of Lot $ | (b) Cost of Improvements $ | Total (a + b) $

Complete this line if this is a refinance loan.

Year Acquired | Original Cost $ | Amount Existing Liens $ | Purpose of Refinance | Describe Improvements ☐ made ☐ to be made Cost: $

Title will be held in what Name(s) | Manner in which Title will be held | Estate will be held in: ☐ Fee Simple ☐ Leasehold (show expiration date)

Source of Down Payment, Settlement Charges and/or Subordinate Financing (explain)

III. BORROWER INFORMATION

Borrower	Co-Borrower
Borrower's Name (include Jr. or Sr. if applicable)	Co-Borrower's Name (include Jr. or Sr. if applicable)
Social Security Number / Home Phone (incl. area code) / Age / Yrs. School	Social Security Number / Home Phone (incl. area code) / Age / Yrs. School
☐ Married ☐ Unmarried (include single, divorced, widowed) ☐ Separated / Dependents (not listed by Co-Borrower) no. ages	☐ Married ☐ Unmarried (include single, divorced, widowed) ☐ Separated / Dependents (not listed by Borrower) no. ages
Present Address (street, city, state, ZIP) ☐ Own ☐ Rent ____ No. Yrs.	Present Address (street, city, state, ZIP) ☐ Own ☐ Rent ____ No. Yrs.

If residing at present address for less than two years, complete the following:

Borrower	Co-Borrower
Former Address (street, city, state, ZIP) ☐ Own ☐ Rent ____ No. Yrs.	Former Address (street, city, state, ZIP) ☐ Own ☐ Rent ____ No. Yrs.
Former Address (street, city, state, ZIP) ☐ Own ☐ Rent ____ No. Yrs.	Former Address (street, city, state, ZIP) ☐ Own ☐ Rent ____ No. Yrs.

IV. EMPLOYMENT INFORMATION

Borrower	Co-Borrower
Name & Address of Employer ☐ Self Employed / Yrs. on this job / Yrs. employed in this line of work/profession	Name & Address of Employer ☐ Self Employed / Yrs. on this job / Yrs. employed in this line of work/profession
Position/Title/Type of Business / Business Phone (incl. area code)	Position/Title/Type of Business / Business Phone (incl. area code)

If employed in current position for less than two years or if currently employed in more than one position, complete the following:

Borrower	Co-Borrower
Name & Address of Employer ☐ Self Employed / Dates (from - to) / Monthly Income $	Name & Address of Employer ☐ Self Employed / Dates (from - to) / Monthly Income $
Position/Title/Type of Business / Business Phone (incl. area code)	Position/Title/Type of Business / Business Phone (incl. area code)
Name & Address of Employer ☐ Self Employed / Dates (from - to) / Monthly Income $	Name & Address of Employer ☐ Self Employed / Dates (from - to) / Monthly Income $
Position/Title/Type of Business / Business Phone (incl. area code)	Position/Title/Type of Business / Business Phone (incl. area code)

Freddie Mac Form 65 10/92 Fannie Mae Form 1003 10/92

VMP-21 (9210) Page 1 of 4 Printed on Recycled Paper

VMP MORTGAGE FORMS • (313)293-8100 • (800)521-7291

V. MONTHLY INCOME AND COMBINED HOUSING EXPENSE INFORMATION

Gross Monthly Income	Borrower	Co-Borrower	Total	Combined Monthly Housing Expense	Present	Proposed
Base Empl. Income*	$	$	$	Rent	$	
Overtime				First Mortgage (P&I)		$
Bonuses				Other Financing (P&I)		
Commissions				Hazard Insurance		
Dividends/Interest				Real Estate Taxes		
Net Rental Income				Mortgage Insurance		
Other (before completing, see the notice in "describe other income," below)				Homeowner Assn. Dues		
				Other:		
Total	$	$	$	**Total**	$	$

* Self Employed Borrower(s) may be required to provide additional documentation such as tax returns and financial statements.

B/C	Describe Other Income *Notice:* Alimony, child support, or separate maintenance income need not be revealed if the Borrower (B) or Co-Borrower (C) does not choose to have it considered for repaying this loan.	Monthly Amount
		$

VI. ASSETS AND LIABILITIES

This Statement and any applicable supporting schedules may be completed jointly by both married and unmarried Co-Borrowers if their assets and liabilities are sufficiently joined so that the Statement can be meaningfully and fairly presented on a combined basis; otherwise separate Statements and Schedules are required. If the Co-Borrower section was completed about a spouse, this Statement and supporting schedules must be completed about that spouse also.

Completed ☐ Jointly ☐ Not Jointly

ASSETS Description	Cash or Market Value
Cash deposit toward purchase held by:	$
List checking and savings accounts below	
Name and address of Bank, S&L, or Credit Union	
Acct. no.	$
Name and address of Bank, S&L, or Credit Union	
Acct. no.	$
Name and address of Bank, S&L, or Credit Union	
Acct. no.	$
Name and address of Bank, S&L, or Credit Union	
Acct. no.	$
Stocks & Bonds (Company name/number & description)	$
Life insurance net cash value	$
Face amount: $	
Subtotal Liquid Assets	$
Real estate owned (enter market value from schedule of real estate owned)	$
Vested interest in retirement fund	$
Net worth of business(es) owned (attach financial statement)	$
Automobiles owned (make and year)	$
Other Assets (itemize)	$
Total Assets a.	$

Liabilities and Pledged Assets. List the creditor's name, address and account number for all outstanding debts, including automobile loans, revolving charge accounts, real estate loans, alimony, child support, stock pledges, etc. Use continuation sheet, if necessary. Indicate by (*) those liabilities which will be satisfied upon sale of real estate owned or upon refinancing of the subject property.

LIABILITIES	Monthly Pmt. & Mos. Left to Pay	Unpaid Balance
Name and address of Company Acct. no.	$ Pmt./Mos.	$
Name and address of Company Acct. no.	$ Pmt./Mos.	$
Name and address of Company Acct. no.	$ Pmt./Mos.	$
Name and address of Company Acct. no.	$ Pmt./Mos.	$
Name and address of Company Acct. no.	$ Pmt./Mos.	$
Name and address of Company Acct. no.	$ Pmt./Mos.	$
Name and address of Company Acct. no.	$ Pmt./Mos.	$
Alimony/Child Support/Separate Maintenance Payments Owed to:	$	
Job Related Expense (child care, union dues, etc.)	$	
Total Monthly Payments	$	
Net Worth (a minus b) ➧ $	**Total Liabilities b.**	$

Freddie Mac Form 65 10/92 Page 2 of 4 Fannie Mae Form 1003 10/92

VI. ASSETS AND LIABILITIES (cont.)

Schedule of Real Estate Owned (If additional properties are owned, use continuation sheet.)

Property Address (enter S if sold, PS if pending sale or R if rental being held for income)		Type of Property	Present Market Value	Amount of Mortgages & Liens	Gross Rental Income	Mortgage Payments	Insurance, Maintenance, Taxes & Misc.	Net Rental Income
			$	$	$	$	$	$
		Totals	$	$	$	$	$	$

List any additional names under which credit has previously been received and indicate appropriate creditor name(s) and account number(s):

Alternate Name	Creditor Name	Account Number

VII. DETAILS OF TRANSACTION

a. Purchase price	$
b. Alterations, improvements, repairs	
c. Land (if acquired separately)	
d. Refinance (incl. debts to be paid off)	
e. Estimated prepaid items	
f. Estimated closing costs	
g. PMI, MIP, Funding Fee	
h. Discount (if Borrower will pay)	
i. Total Costs (add items a through h)	
j. Subordinate financing	
k. Borrower's closing costs paid by Seller	
l. Other Credits (explain)	
m. Loan amount (exclude PMI, MIP, Funding Fee financed)	
n. PMI, MIP, Funding Fee financed	
o. Loan amount (add m & n)	
p. Cash from/to Borrower (subtract j, k, l & o from i)	

VIII. DECLARATIONS

If you answer "Yes" to any questions a through i, please use continuation sheet for explanation.	Borrower Yes	Borrower No	Co-Borrower Yes	Co-Borrower No
a. Are there any outstanding judgments against you?	☐	☐	☐	☐
b. Have you been declared bankrupt within the past 7 years?	☐	☐	☐	☐
c. Have you had property foreclosed upon or given title or deed in lieu thereof in the last 7 years?	☐	☐	☐	☐
d. Are you a party to a lawsuit?	☐	☐	☐	☐
e. Have you directly or indirectly been obligated on any loan which resulted in foreclosure, transfer of title in lieu of foreclosure, or judgment? (This would include such loans as home mortgage loans, SBA loans, home improvement loans, educational loans, manufactured (mobile) home loans, any mortgage, financial obligation, bond, or loan guarantee. If "Yes," provide details, including date, name and address of Lender, FHA or V.A. case number, if any, and reasons for the action.)	☐	☐	☐	☐
f. Are you presently delinquent or in default on any Federal debt or any other loan, mortgage, financial obligation, bond, or loan guarantee? If "Yes," give details as described in the preceding question.	☐	☐	☐	☐
g. Are you obligated to pay alimony, child support, or separate maintenance?	☐	☐	☐	☐
h. Is any part of the down payment borrowed?	☐	☐	☐	☐
i. Are you a co-maker or endorser on a note?	☐	☐	☐	☐
j. Are you a U.S. citizen?	☐	☐	☐	☐
k. Are you a permanent resident alien?	☐	☐	☐	☐
l. **Do you intend to occupy the property as your primary residence?** If "Yes," complete question m below.	☐	☐	☐	☐
m. Have you had an ownership interest in a property in the last three years?	☐	☐	☐	☐
(1) What type of property did you own–principal residence (PR), second home (SH), or investment property (IP)?				
(2) How did you hold title to the home–solely by yourself (S), jointly with your spouse (SP), or jointly with another person (O)?				

IX. ACKNOWLEDGMENT AND AGREEMENT

The undersigned specifically acknowledge(s) and agree(s) that: (1) the loan requested by this application will be secured by a first mortgage or deed of trust on the property described herein; (2) the property will not be used for any illegal or prohibited purpose or use; (3) all statements made in this application are made for the purpose of obtaining the loan indicated herein; (4) occupation of the property will be as indicated above; (5) verification or reverification of any information contained in the application may be made at any time by the Lender, its agents, successors and assigns, either directly or through a credit reporting agency, from any source named in this application, and the original copy of this application will be retained by the Lender, even if the loan is not approved; (6) the Lender, its agents, successors and assigns will rely on the information contained in the application and I/we have a continuing obligation to amend and/or supplement the information provided in this application if any of the material facts which I/we have represented herein should change prior to closing; (7) in the event my/our payments on the loan indicated in this application become delinquent, the Lender, its agents, successors and assigns, may, in addition to all their other rights and remedies, report my/our name(s) and account information to a credit reporting agency; (8) ownership of the loan may be transferred to successor or assign of the Lender without notice to me and/or the administration of the loan account may be transferred to an agent, successor or assign of the Lender with prior notice to me; (9) the Lender, its agents, successors and assigns make no representations or warranties, express or implied, to the Borrower(s) regarding the property, the condition of the property, or the value of the property.

Certification: I/We certify that the information provided in this application is true and correct as of the date set forth opposite my/our signature(s) on this application and acknowledge my/our understanding that any intentional or negligent misrepresentation(s) of the information contained in this application may result in civil liability and/or criminal penalties including, but not limited to, fine or imprisonment or both under the provisions of Title 18, United States Code, Section 1001, et seq. and liability for monetary damages to the Lender, its agents, successors and assigns, insurers and any other person who may suffer any loss due to reliance upon any misrepresentation which I/we have made on this application.

Borrower's Signature	Date	Co-Borrower's Signature	Date
X		X	

X. INFORMATION FOR GOVERNMENT MONITORING PURPOSES

The following information is requested by the Federal Government for certain types of loans related to a dwelling, in order to monitor the Lender's compliance with equal credit opportunity, fair housing and home mortgage disclosure laws. You are not required to furnish this information, but are encouraged to do so. The law provides that a Lender may neither discriminate on the basis of this information, nor on whether you choose to furnish it. However, if you choose not to furnish it, under Federal regulations this Lender is required to note race and sex on the basis of visual observation or surname. If you do not wish to furnish the above information, please check the box below. (Lender must review the above material to assure that the disclosures satisfy all requirements to which the Lender is subject under applicable state law for the particular type of loan applied for.)

BORROWER

☐ I do not wish to furnish this information

Race/National Origin: ☐ American Indian or Alaskan Native ☐ Asian or Pacific Islander ☐ White, not of Hispanic Origin ☐ Black, not of Hispanic origin ☐ Hispanic ☐ Other (specify) ____________

Sex: ☐ Female ☐ Male

CO-BORROWER

☐ I do not wish to furnish this information

Race/National Origin: ☐ American Indian or Alaskan Native ☐ Asian or Pacific Islander ☐ White, not of Hispanic Origin ☐ Black, not of Hispanic origin ☐ Hispanic ☐ Other (specify) ____________

Sex: ☐ Female ☐ Male

To be Completed by Interviewer	Interviewer's Name (print or type)	Name and Address of Interviewer's Employer
This application was taken by: ☐ face-to-face interview ☐ by mail ☐ by telephone	Interviewer's Signature / Date	
	Interviewer's Phone Number (incl. area code)	

Chapter 4

Cost Estimating

ESTIMATING THE COST of a new house is not an exact science. You can, however, come reasonably close to estimating a total cost and you can guide your construction project toward that estimated cost. You can do this only if *you* are the general contractor.

There is a catch-22 in building a home with regard to estimating costs. As you'll see, you can't get a really accurate estimate of costs until you have house plans, and you can't — or shouldn't — get house plans until you know how much a certain style and size home will cost to build in your area of the country. If you spend the money to buy plans and then find that it's too expensive to build that particular plan, you've wasted time and money.

Well, fear not, for I've invented a process for determining how much it costs to build a house anywhere in the United States or Canada. I call this process "windshield estimating." It is called this because you can literally determine building costs from the comfort of your car as you drive around your area and find the style and size home you're interested in. Visit areas where quality custom homes are being built by small to medium-size builders, for this is your ultimate goal — a quality custom home by a very small builder (you). Find a style of home you like and a size you find adequate. Forget the selling price of the houses you are looking at, since your only concern is to determine building costs.

Let's say you find a nice two-story house, with a three-car garage and a full, unfinished basement listed for sale at $XX.00. All you have to do then is subtract 25 percent of the selling price, then subtract the property cost, and you are left with the actual cost of building that house. Divide that final cost figure by the square footage of living space and you've got the cost per square foot in your area. Simple?

You bet! Does this really work? Yes, for most areas. Just be sure you look at as many houses as you can to get a good feel for what is accurate and fair pricing in your area. If you live in a rural area and not much is being built, travel to the nearest town or buy the newspaper and start looking.

Now, there are two questions that should have come to mind immediately when you read the above. One, how do you know the property cost? And two, how do you know the square footage of the house you're looking at so that you can divide and get cost per square foot? In answer to question one, the property cost can be determined by checking the selling price of adjoining property or nearby comparable property, contacting a realtor, calling the local tax office (all real estate transactions and records are always a matter of public record), or checking with the builder.

The builder is the most obvious source of information, as he or she will be interested in your future business and would therefore have to tell you not only the lot value, but — guess what! — the square footage as well. If you can't talk to the builder to find out square footage, call the building inspection department and ask (as it too is a matter of public record) the square footage that was recorded when the building permit was issued. At any rate, with a bit of perseverence, you'll come up with building costs on a square foot basis without a lot of effort.

Now you can determine what style and size house plan to start looking for. And, once you get your plans, you can then do a more precise estimate. For more precise estimating, draw up an estimate sheet like the one on page 45, and fill in each item based on bids from subcontractors and suppliers. Enter these bids under the *estimated cost* column. The *actual cost* column is for entering paid bills. This column will be accurate because it will be filled in when each category is completed and has been paid for.

Most lumber companies will do a take-off (estimate) of the materials they hope to sell you. For your first house I'd advise you to let them. That doesn't mean you have to buy from them. More on this in Chapter 7. Your carpenter can also do a take-off of materials he will use. He may or may not charge for this service. Ask him.

As you pay for items during construction, compare them to your original estimate so you can see how the cost is running. If costs run higher than estimated in some categories, you may want to trim some costs in others.

Most of the heavy costs come at the beginning of construction for such items as lumber, masonry, carpentry, plumbing, electrical work, and heating plants. Some of the items bought later are still quite significant in affecting the total cost of the house. These include carpeting, appliances, bath accessories, wallpapering (which can be postponed, to save money), parquet floors and other flooring, and special *moldings*. It is possible to cut costs on these.

I have seen houses with $5,000 worth of molding just on the first floor. Molding can be added later if reducing costs is important. Windows can vary as much as $100 each just because of brand name, and the difference in quality is negligible. Do you need a ten-cycle dishwasher, when a five-cycle or a two-cycle will do as well? So it is with each item.

Balancing Costs

If, as you prepare your estimate using the form on page 45, you figure $8,000 for framing lumber and supplies, and the actual cost turns out to be $8,500, start looking for other places to trim. If the actual cost turns out to be $7,500, don't become too elated and start looking for ways to spend that $500. It will disappear before you finish, I promise.

Why would a single item vary so much? For three reasons:

1. Prices of materials vary daily — sometimes substantially.
2. Different tradespeople use different amounts of materials. Some carpenters overbuild and will use $300 to $400 worth of more lumber to do the same job as someone else. (Let them. I'd rather overbuild in framing than underbuild.) This is just one house, and you can save somewhere else. If you were building ten to fifty houses a year, you would need to be more careful with materials used.
3. It is impossible to determine the exact number of *studs*, *joists*, bricks, and other materials that go into a house. Some estimators are more accurate than others, but an exact estimate is pure luck.

Don't worry yourself to death over the lack of exactness. A builder pads an estimate by at least 5 percent to cover these contingencies.

COST ESTIMATE

Owner's Name ______________________________

Property Address ______________________________

City ______________________________

	ESTIMATED COST	ACTUAL COST
Permits, fees, surveys, etc. ______	______	______
Utilities (electric, gas, phone) ______	______	______
Excavation ______	______	______
Foundation ______	______	______
Rough Lumber ______	______	______
Rough Labor ______	______	______
Windows & Exterior Doors ______	______	______
Roofing ______	______	______
Concrete Flatwork ______	______	______
Siding ______	______	______
Plumbing ______	______	______
Heating ______	______	______
Electrical ______	______	______
Insulation ______	______	______
Water (Well) ______	______	______
Sewer (Septic) ______	______	______
Fireplaces ______	______	______
Drywall ______	______	______
Cabinets ______	______	______
Interior Trim ______	______	______
Interior Trim Labor ______	______	______
Painting ______	______	______
Appliances ______	______	______
Light Fixtures ______	______	______
Floor Coverings ______	______	______
Driveway ______	______	______
Garage Door ______	______	______
Other ______	______	______
Subtotal ______	______	______
Misc. (5% to 10% recommended) ______	______	______
Total Building Cost ______	______	______
Land Cost ______	______	______

Total Project Cost (Land & Building Costs) ______________

Downpayment ______________________________

Construction Loan Amount ______________________

That is just one reason why you'll be saving by being your own general contractor. Don't worry about that lack of exactness, but keep an eye on each category as you get bids and prices.

The number of bricks and concrete blocks required is the most difficult to estimate. Brick companies often will estimate for you, but they won't be any more accurate than you on your first try.

You have a choice of making a rough estimate and ordering from it, or having the brick company do it. If you don't have enough bricks, you can order more. If you order too many, you can return the surplus. (Ask the brick company about its return policy when you're shopping for your brick.) Or, quite often, your masonry sub will buy unused brick at a discounted price and haul them off the job for you. Ordering bricks may be the most difficult part of estimating — but only if you let it bother you. I recommend ordering a little short and letting the brick company hold some extra bricks in its brickyard in your name so that if you need to order more, you'll get the same *color run.*

You'll need about five to seven bags of mortar cement per 1,000 bricks. Sand is ordered in such enormous quantities — fourteen or sixteen cubic yards — that you don't have to concern yourself with exactness. If I seem to oversimplify a very complex item — good. Don't worry about it. Order a little less than you think you need of sand, brick, and block, and if you run out, order more. Your masonry sub will be sure to let you know in advance if he is going to run short of an item, because if he doesn't he will be held up.

Many masonry subs can estimate needs. If you find one who does, let him. Some even supply their own materials — better yet.

Never be too proud to ask questions, from either a supplier or subcontractor. Both are very willing to help. They don't make any money until they sell you something or perform a service. Also, some plans purchased in magazines come with — or make available for a fee — a schedule of materials, right down to the amount of nails. They are fairly accurate and I wouldn't hesitate to use them. Suppliers also do this for free. (See Chapter 7 for types of suppliers to approach.)

Cost Breakdowns

Here are the items and categories you will most likely encounter in building your home. With some, accurate bids from suppliers or sub-

Estimating Bricks Needed

To figure the area of a wall, multiply the length times the width (or the base times the height). If the wall is to be 8 feet high and the base is 20 feet long, the area is 160 square feet.

If you plan a brick foundation one *course* in thickness, which is what you would have on a house with a relatively low crawl space, here is how to determine how many standard 2" × 4" × 8" bricks you would need.

The foundation is 30' × 60' × 3'. Treat each wall as a separate area, determine that area, add all areas together, and multiply by eight. Multiplying by 8 is generous because it requires 7.5 bricks plus mortar space per square foot of surface area.

So our 30' × 60' × 3' foundation would use the following:

Side 1	30 × 3 = 90 square feet
Side 2	60 × 3 = 180 square feet
Side 3	30 × 3 = 90 square feet
Side 4	60 × 3 = 180 square feet
Total	540 square feet × 8 = 4,320 bricks

contractors are possible, but with others, a guesstimated allowance can be used.

- **Permits, fees, surveys.** Your building inspection department can give you the cost of permits and fees, which vary by locale. Your lender will require you to have a mortgage survey and you will also have to insure your building project against loss due to fire or other acts of God. Any other costs necessary to get your project started can be included in this category and some guesstimating can be utilized.

- **Utilities (electric, gas, phone).** In some rural areas this item can be very costly, running into the thousands of dollars. Check with your local utility companies in advance.

- **Excavation.** This item will depend on such factors as locale, terrain, season of the year, and soil. Get a written bid if possible.
- **Foundation.** This item will vary considerably based on factors such as slope of lot, and the height of basement walls.
- **Rough Lumber.** A list on page 82 gives you an idea of what goes into the "framing package" as it is called. This includes all materials except windows, doors, and roof shingles (although it can include those as well). These are all the materials necessary to *dry-in* the house. A good lumber company will put together this material list for you free of charge because they want your business.
- **Rough Labor.** This is the labor required to bring the house to the dry-in stage. After this stage is completed, all other stages can commence, some simultaneously. The best way to contract for this job is by the square foot, with the square footage agreed to before you start. Five people will arrive at five different square footage totals, using the same set of plans. Some will vary by 300 square feet or more. Sounds incredible, doesn't it? But I swear that it's true. In determining the square footage, houses are measured from outside wall to outside wall, *not* from roof overhangs. If the house is not easily divided into rectangles for simplifying square footage determination and you can't figure it out, have the designer do it for you. Ready-made plans generally come with the square footage broken down for you. Use those figures.
- **Windows and Exterior Doors.** This cost is simple to estimate since you have an exact count. I do not recommend any particular brand, but I do recommend that you visit a couple of building supply companies and compare. Most carry more than one brand. Locally made windows usually are less expensive than national brands and give as good warranty service. For special windows such as angular or bay windows get exact quotes from the supplier. Generally there is no additional cost to install windows (except special or unusual ones) as that labor is included in your carpenter's framing charge. Be sure it is.

Exterior doors are as easy to estimate as windows, and all the same factors apply, including the labor to install being included in the framing bill from your carpenter. Sliding glass doors which will be slightly more expensive and may require a separate installation charge, depending on your carpenter. Be sure to ask whoever installs sliding glass doors to caulk under the threshold thoroughly, preferably with silicone type caulking.

- **Roofing.** It is measured and estimated by squares. A square of roofing is 10 feet square or 100 square feet. The number of squares can be determined by your supplier, your carpenter, or you. I advise either of the first two for your first house. They won't be exact, but may come a little closer than you will. Shingles are priced by their weight per square and the material from which they are made. Generally we are talking about either asphalt or fiberglass and their weight per square is from 245 pounds to 345 pounds. Also there are cedar shakes and slate, tin, tar, and gravel (built up roofs). But in this book we will deal with asphalt and fiberglass. The cost of labor to install will depend upon the sub, the weight of the shingle, and the pitch of the roof — the steeper the pitch the higher the price for labor. The same applies to weight. The heavier they are, the higher the price. Most roofs are the 245-pound asphalt variety and most roof pitches average about six inches for each foot of horizontal travel (a 6/12 pitch).

- **Concrete flatwork** (slabs), garage floors, basement floors. This refers to smooth finish concrete work, not rough finish as in driveways, patios, and walks. It also involves the use of other materials such as Styrofoam, wire mesh, *expansion joints* and *polyurethane.* Your concrete subcontractor can explain this to you. The work is closely inspected by most building departments. Get a bid based on square footage of actual concrete area.

- **Siding.** Depending on the material used (which may range from brick to vinyl), you can get an accurate bid from a subcontractor that includes the cost of materials.

- **Plumbing.** This bid should include all fixtures such as water closets, sinks, and water heater. It will *not* include such items as dishwasher, garbage disposal, washing machine, or other household appliances.
- **Heating.** Use heating and air conditioning systems recommended by your local utility company experts. Cost of installation should include all other ventwork such as bathrooms, stoves, and clothes dryer.
- **Electrical.** In addition to wiring costs, this bid should include all switches, receptables, wires, panels and breakers, wiring of all built-in appliances, and compliance with codes.
- **Insulation.** Get a bid per local code for minimum insulation. To get the maximum insulation for number of dollars spent, consult your local utility company experts.
- **Water (well).** For water tap-in fees call your local municipality. For a well bid, call a well drilling firm familiar with the area and get a firm maximum bid as well as a drilling price per linear foot drilled.
- **Sewer (septic).** For a sewer tap-in fee, call your local municipality. For the price of a septic field, get a written bid from a local contractor who does septic system installation.
- **Fireplaces.** Whether it is masonry or prefab, get a bid in writing.
- **Drywall.** Bids should include labor and materials to hang, tape joints, and finish (two coats).
- **Cabinets.** Bids should include kitchen cabinets and bathroom vanity cabinets. Labor to install should be included in trim labor.
- **Interior Trim.** A bid from the lumber supplier should include all interior doors, mouldings, closet shelves, and stairway trim. It should also include carpet underlayment.
- **Interior Trim Labor.** You should get a bid to install all the materials for both Cabinets and Interior Trim, above.

- **Painting.** Although people usually skimp on this category, often planning to do it themselves, it is best to get an estimate, even if just for the materials. If you are *not* planning on doing the labor yourself, get a written bid for both labor and materials.
- **Appliances.** In the planning stage, you just need to have an idea of what you want in the way of appliances and insert a dollar amount allowance in your estimate. At this stage you don't need the actual models picked out, just a ballpark idea of what you think you will eventually buy. Use a dollar allowance you feel is adequate to get the appliances you want.
- **Light Fixtures.** Though you probably won't have your actual fixtures selected, figure an amount that will be enough to cover the costs of all necessary fixtures.
- **Floor Coverings.** Estimate approximate costs for all floor coverings, taking into account varying amounts for wood, carpet, tile, or other coverings.
- **Driveway.** Depending on the material used, get a bid in writing based on a per square foot of surface area. Make sure you agree on the square footage.
- **Garage Door.** If you can still afford one, get a price from your local lumber company — with or without operators, installed or uninstalled.
- **Other.** Anything not included above, i.e., fences, sidewalks, decks, swimming pools, saunas, etc.

Chapter 5

Further Preparations

YOU HAVE PURCHASED a lot, selected house plans, estimated costs, and arranged your construction loan and permanent mortgage. Several details still need your attention before you break ground.

Obtain a building code book or pamphlet from your local building inspector, listed in the phone book. If you don't have a building inspection department, and many small towns don't, find out about any local or state codes, and how they are enforced. This information can be obtained through City Hall, the Town Hall, your lender, or an attorney. You need not be an expert on codes, but it is best to be familiar with a few of the important specifications, such as those that relate to the thickness of footings, the preparation of concrete slabs, and requirements for foundation walls and lumber spans. If you don't have a building inspection department, a local library should have a code book for your state and/or county, and several good reference books on each trade involved in the construction business. You may want to do this before getting a loan. Regulations could change your plans — and increase your costs.

Again, I emphasize that you need not be an expert. Your subs will know — or should know — the codes as they apply to their particular fields. If there is a question, you'll know where to find the correct answer if you have a code book and reference books. If you find good subs, you may never use those references.

Also, at this point, you must find out the procedure for obtaining your building permit. This can be done over the phone to the building inspection department. Officials will also tell you the procedures for inspections such as when to call for them, and which phases of

BUILDING PERMIT
DEPARTMENT OF BUILDINGS AND SAFETY ENGINEERING

PERMIT NO.

THE BUREAU OF LICENSES AND PERMITS DATE 19

HEREBY GRANTS PERMISSION TO

NAME ADDRESS CONTRACTOR'S STATE LICENSE

TO ( ) STORY NUMBER OF APARTMENTS

ON THE SIDE OF (BUILDING NO.) ST. OR AVE. ZONING DISTRICT

LOT NO. AND SUBDIVISION SIZE

BUILDING IS TO BE FT. WIDE BYFT. LONG BYFT. IN HEIGHT
AND SHALL CONFORM IN CONSTRUCTION TO TYPE

USE GROUP BASEMENT WALLS OR FOUNDATION

REMARKS:

APPROVED BY

CUBIC FEET ESTIMATED COST $ REGULAR FEES $

ZONING FEE $

OWNER BUREAU OF LICENSES AND PERMITS
ADDRESS BY

THIS PERMIT CONVEYS NO RIGHT TO OCCUPY ANY STREET, ALLEY OR SIDEWALK OR ANY PART THEREOF, EITHER TEMPORARILY OR PERMANENTLY. ENCROACHMENTS ON PUBLIC PROPERTY, NOT SPECIFICALLY PERMITTED UNDER THE BUILDING CODE, MUST BE APPROVED BY THE COMMON COUNCIL, STREET OR ALLEY GRADES AS WELL AS DEPTH AND LOCATION OF PUBLIC SEWERS MAY BE OBTAINED FROM THE DEPARTMENT OF PUBLIC WORKS — CITY ENGINEERS OFFICE. THE ISSUANCE OF THIS PERMIT DOES NOT RELEASE THE

APPLICANT FROM THE CONDITIONS OF ANY APPLICABLE SUBDIVISION RESTRICTIONS.

MINIMUM OF THREE CALL INSPECTIONS REQUIRED FOR ALL CONSTRUCTION WORK:

1.FOUNDATIONS OR FOOTINGS.
2.PRIOR TO COVERING STRUCTURAL MEMBERS (READY TO LATH).
3.FINAL INSPECTION FOR COMPLIANCE PRIOR TO OBTAINING CERTIFICATE OF OCCUPANCY.

APPROVED PLANS **MUST** BE RETAINED ON JOB AND THIS CARD **KEPT POSTED** UNTIL FINAL INSPECTION HAS BEEN MADE. WHERE A CERTIFICATE OF OCCUPANCY IS REQUIRED, SUCH BUILDING **SHALL NOT BE OCCUPIED** UNTIL **FINAL INSPECTION** HAS BEEN MADE AND CERTIFICATE OBTAINED.

SEPARATE PERMITS REQUIRED FOR ELECTRICAL AND PLUMBING INSTALLATIONS

POST THIS CARD

BUILDING INSPECTION APPROVALS

1
DRAIN TILE AND FOUNDATION ______
DATE ______
INSPECTOR ______

2
SUPERSTRUCTURE ______
(PRIOR TO LATH AND PLASTER)
DATE ______
INSPECTOR ______

3
FINAL INSPECTION ______
DATE ______
INSPECTOR ______

WORK SHALL NOT PROCEED UNTIL EACH BUREAU HAS APPROVED THEVARIOUS STAGES OF CONSTRUCTION.

ELECTRICAL INSPECTION APPROVALS

1
ROUGHING IN ______
DATE ______
INSPECTOR ______

2
FINAL INSPECTION ______
INSPECTOR ______

SAFETY ENGINEERING APPROVAL
3
APPROVED ______
DATE ______
INSPECTOR ______

INSPECTIONS INDICATED ON THIS CARD CAN BE ARRANGED FOR BY TELEPHONE OR WRITTEN NOTIFICATION.

PLUMBING INSPECTION APPROVALS

1
BUILDING SEWER
(A) SANITARY ______
DATE ______
INSPECTOR ______
(B) STORM ______
DATE ______
INSPECTOR ______

2
CROCK TO IRON ______
DATE ______
INSPECTOR ______

3
UNDERGROUND STORM DRAINS ______
DATE ______
INSPECTOR ______

4
ROUGH PLUMBING ______
DATE ______
INSPECTOR ______

5
WATER PIPING FINAL INSPECTION ______
DATE ______
INSPECTOR ______

construction require inspection. Good subcontractors will have this information and take care of the inspections for you.

Inspections Required

As an example, the county I live in requires these inspections.

- Temporary electrical, or *saw service*, to insure proper grounding.
- Footing, before pouring concrete, to make sure we have reached solid load-bearing ground.
- Slab, before concrete is poured, to determine if it is properly insulated. Any plumbing in the slab is also inspected at this time.
- Electrical, plumbing, heating, and air-conditioning rough-in, to insure that in-the-wall installations that can't be seen later are correct.
- Framing rough-in, to insure structural integrity, especially after electrical, plumbing, heating, and air-conditioning installations. These workers have been known to cut too deeply into a joist or stud and weaken it.
- Insulation, to insure compliance with new local standards.
- Final electrical, plumbing, heating, and air-conditioning and building to insure they work properly, comply with codes, and are safe. All final inspections must be completed before we can get more than temporary electrical service.

You can now select materials such as brick, shingles, windows, doors, roofing, trim, plumbing, fixtures, built-in appliances, light fixtures, flooring, hardware, and wallpaper. As you visit building supply houses to see their samples, you can also take care of a couple of other important tasks:

- Open your line of credit with them. This is quite easy, even for an inexperienced builder. See Chapter 7 for a list of the types of suppliers you'll need.

- Ask them to recommend reliable subcontractors. Smaller building supply companies are better equipped to furnish this information because the management is more directly involved with customers. The key sub you are looking for is your carpenter, and most of the good ones patronize building supply houses from time to time. Many make building supply houses their source of referrals. This is especially true in resort areas. (Don't forget, you can also use this book as a guide to building a second home in a resort area.) You may find other subs here. More in the next chapter about subs.

Now is the time to contact your gas, electric, and water departments for hookup procedures. Every locale has different utilities and thus different procedures. If you so desire, you will also want to order a *prewire* from your phone company at this time as the company sometimes requires three to five weeks' notice. Some don't charge for this, some do. Ask first.

You will also want to shop around for a builders' risk or fire policy from various insurance agents. In most states, agents are fairly equal in policy offerings, since they're governed by state laws. But be sure to check. The policy should take effect when materials arrive on the job but your lender may want it to be in force before closing the construction loan. The lender will be the payee in the policy since it is the company's money. When the house is complete, and you move in, the fire policy can be converted to homeowners' insurance, often at a fairly good savings over a new homeowners' policy.

Doing Your Own Labor

If you plan to do any of your own labor, I have only one thing to say: if you aren't an expert at the particular trade you plan to do yourself, forget it. This is especially true in painting. People think it is so easy, but unfortunately it comes at a time when the house is quite far along, and construction disbursements will be approaching their maximum amount. You need painters who can get in and get out, do the job fast, and at the same time do a respectable job. If you seek perfection in painting, wait until you are living in the house and do your own touch-up painting.

If you hire painters by the hour, plan to stay with them the whole time, or else plan on it taking twice as long. You won't get a better job, or at least that much better, for the extra time. Hire them by a contract amount, such as ninety cents a square foot based on the amount of heated area.

When it comes to doing your own work, remember one of Murphy's laws — "Nothing is as easy as it looks."

Chapter 6

Subcontractors

A SUBCONTRACTOR (SUB) IS an individual or a firm that contracts to perform part or all of another contract. In your case you are technically the builder or general contractor, and you will build by subcontracting with others. You will pay for this by a predetermined contract amount with each one. This is important. You will have no hourly wage employees working for you. Thus you will avoid all of the governmental red tape and taxes concerning employees. Your subcontractors are not considered to be employees.

Your Subcontractors, Professionals

Here is a list of the subcontractors and professional people you probably will be contracting with, listed generally in the order in which you will need them.

- Attorney. Needed for initial assistance for questions, land purchase, loan assistance, *loan closing*, continuing assistance if necessary for legal questions and/or settling disputes with subs or suppliers. I would not attempt the project without an attorney. You can find one who specializes in real estate by calling the Lawyer's Referral Services (in the Yellow Pages) or a real estate company, or asking a friend. Attorney's fees are a cost of construction.
- Lending officers at savings and loans, banks, or mortgage companies.

- House designer or architect.
- Carpentry sub, your key sub, to be lined up early.
- Surveyor.
- Grading and excavation contractor.
- Footing contractor.
- Brick masonry contractor. Builds the foundation.
- Concrete (finisher) subcontractor. Pours the slab or concrete floor.
- Waterproofing contractor.
- Electrical contractor.
- Plumbing contractor (and well and septic system, if needed).
- Heating and air-conditioning and vent (HVAC) contractor.
- Roofing contractor.
- Insulation contractor.
- Drywall contractor.
- Painting contractor.
- Flooring, carpet, and Formica contractor.
- Tile contractor.
- Cleaning crew contractor.
- Landscape contractor.

Finding Your Subs

As mentioned earlier, a lumber supply store is the best place to start for finding your carpenter subcontractor. Your carpenter sub will be able to recommend almost everyone else, as he is on the job more than anyone else and knows most of the other subs involved in building a house.

Carpentry Labor Contract

To: (your name)__________ Sub cont. ______________
(address) ____________________________________

____

Date: ________________ Job address ____________

____

Owner: ________________

Area: Heated ______________ sq.ft.
Unheated ____________ sq.ft.
Decks ______________ sq.ft.

Charges

Framing ______________ @__sq.ft.x__sq.ft. = $ ________
Boxing and siding _______ @__sq.ft.x__sq.ft. = $ ________
Interior trim ___________ @__sq.ft.x__sq.ft. = $ ________
Decks _______________ @__sq.ft.x__sq.ft. = $ ________
Setting Fireplace $ ______
Setting Cabinets $ ______
Paneling $ ______
Misc. $ ______
Total Charges $ ______
Signed: (your name) __________________ Date: ______
Signed: (subcontractor) ________________ Date: ______

"A good sub is a working sub," especially during a recession. This is not always true, but is a pretty safe bet. The really good ones are sought after and in demand because they do good work and are reliable.

If you can't find a sub through a supplier or your carpenter, the next best place to look is on a job. Find a house under construction. Stop and ask around. You can get names, prices, and references. This takes only a few minutes. It is done all the time and the general contractor shouldn't mind. He probably won't even be there. Besides, he does (or his superintendents do) the very same thing.

Often the boss or owner of a subcontracting firm is on the job. Get his number and arrange a meeting. Sometimes there are signs at the job site advertising different subs.

Subs of only certain trades are listed in the Yellow Pages. Carpenters and most independent brick masons are not listed. You'll probably find heating and air-conditioning companies, plumbers, electricians, roofers, waterproofing companies, lumber dealers, and appliance manufacturers.

I had what is probably one of the best drywall subcontractors in the Southeast, but he was not listed. He didn't live in the same city; he lived in the country. He was so good that he stayed backlogged two or three weeks even during recessions. His name was given to me by a lumber dealer.

Subcontractors Contract

Subcontractor: ______________________________

Address ______________________________

Builder: ______________________________

Date ______________________ Plan No. __________

W. Comp. Ins. Co. & Agent ______________________

Certificate No.: __________ Expiration Date: _____

Location of Work: ______________________________

Total Price per House: __________ ($......................)Dollars

Terms of Payment: ______________________________

Work to Be Performed and Materials To Be Supplied:

For Subcontractor: ______________________________

Signature and Title

For Builder: ______________________________

Signature and Title

Each subcontractor should carry insurance on his or her employees, and should provide you with a certificate of insurance. (See example in Appendix.)

It should not be a problem if most subs in your area belong to unions. I state this because some people expect union labor to cost more or fear they will have to do excessive paperwork if they hire union members. Not so. Your subcontractor will be responsible for all compliances to the union and the prices should be competitive.

Since this is your first experience and you won't be familiar with prices in your area, get three or four bids, or quotes, before selecting a sub. Use a written contract with all subs.

You may use the very simple ones I have provided or your attorney can provide you with one. Subcontractors may have their own contracts. At any rate, use one. Don't trust anyone's memory when it comes to dollars. Be sure your bids are comparable, that all are bidding on exactly the same work. If your specifications are properly spelled out in your plans, the bidding should be.

Bids for Plumbing

Plumbing bids should include all plumbing fixtures right down to the toilet seat. They will not include accessories such as toilet paper holders. If colored fixtures are to be used, specify color and brand. Plumbing showrooms are your best bet for the selection of these fixtures. Magazines and brochures don't tell you enough, nor give prices. Most plumbing showrooms won't tell you the wholesale price, but you'll be paying list anyway, as the plumber makes a profit on each fixture and it's included in the bid. Don't make an issue of this. The small profit in the fixtures is one of the plumber's sources of income and he earns it.

Your heating and air-conditioning contract should include vents, for the bath fan, clothes dryer, stove, and range hood. Electrical bids should include all switches, wiring, receptacles, circuit breakers, and their respective *panel boxes*, a temporary service box and installation, *saw service*, wiring of all built-in appliances, and installation of ovens and ranges, furnaces, heaters, and air-conditioners.

All subcontractors should be responsible for obtaining inspections from the building department, but you will want to be sure they do or you will have to do it yourself. Lack of inspections can cause delays. Proceeding without getting inspections can be troublesome and expensive. If you, for example, put up drywall before having your wiring inspected, you could be made to tear down some of the drywall for the electrical inspector. This is not likely to happen, but the inspector could force the issue.

Utilities must be connected. Exactly who is responsible for running water lines, sewer lines, and electrical hookups will vary with each subcontractor involved. Get the responsibility pinned down when you are hiring the subs, then follow through to be sure it is done properly.

Scheduling Your Subcontractors

Try to schedule your subs to fit into the sequence of events outlined in Chapter 8. This won't always be possible, and one sub can hold up the process. This is why you should check their references. Reliability is as important as quality and in some cases more so.

Paying Your Subcontractors

When you sign your contract with your carpenter, you will agree on a contract price for the work. It is usually based on X number of dollars per square foot of heated area and X number of dollars per square foot of under roof, such as in the garage. Prices will vary with the area, unions, and the complexity of the job.

Never pay a sub for work not done, for work that is incomplete, or for an unsatisfactory job. Never pay in advance.

You should work out a schedule of payment with your carpenter. The carpenter and some of the other subs may require draws, or partial payments, as work progresses. This should be discussed before work begins. Don't be shy about it. They are accustomed to discussing such matters.

It is all right to pay a draw, but never pay more than for work done. If, for example, your carpenter says he is 50 percent finished framing, but he has only completed the floor framing and wall framing, and hasn't finished the ceiling joists, roof framing, *sheathing* or *bridging*, he isn't 50 percent complete. He is nearer 40 percent complete.

If he is charging $1 per square foot of area under roof (including the garage, storage, and porches) you will owe him $2,400 total when he has finished framing a 2,400 square foot house. At 40 percent complete, you would give him a maximum of 33 percent or $800.

Plumbers and electricians usually get 60 percent of the total contract price when their rough-in has been completed and inspected. Heating, vent, and air-conditioning rough-in payments depend upon the installation of equipment such as furnaces. If payment is just for duct work and some low voltage wiring, 20 percent of the total should suffice. If a furnace had to be installed during rough-in, add another 10 percent. Work out the arrangement with the sub before he starts. Subs almost always would like to get more money than they have in the job. Be sure there is enough money left in the total bid to complete the job if the HVAC dealer goes broke while you are still building. It has happened. You don't want to be stuck paying more to complete his job. You'll be covered better if you don't overpay him on his rough-in.

Brick masons and painters are about the only other subs who will require a draw in progress. You will have to use your best judgment as to how much of the job is done. Again, don't get ahead of them in paying.

I seem to be saying the only way you'll get the job completed is if you owe your sub money. In some cases that is true, but in others it is only partially true. Some subs would finish regardless. The main objective is to get your job finished. Often subs will have more than one job going at one time. You want yours finished quickly and before one that was started after yours.

Make sure inspections by your county or city are completed and the work is approved before you make any payments at any phase of construction, other than partial draws. This is your assurance that the job will be done, and done properly.

Working with Subcontractors

If you get along well with people, you don't need to read the next few paragraphs. But if you have trouble, read them carefully. The fault — sometimes — may be yours.

At this point you've selected your subs. You're satisfied, now that you've checked them out, that they are honest, trustworthy, and experts in their fields.

Now let them work. Don't try to supervise every blow of a hammer, the placement of every stud. They know more about their trades than you do, and probably, if they came to you well recommended, they take pride in their work. Let them do it.

And, more emphatically, don't try to tell your subs their jobs just because you have read this book and a few others. You'll get good work out of your subs if they understand you think they know their jobs, and you're depending on them for good advice and quality work.

Chapter 7

Suppliers

The suppliers that you will be buying from are listed in this chapter by the type of products they sell. Some sell several different products and you can shorten your shopping time when you patronize them. All of them will require a credit check on you by means of the references that you give them. Usually three charge accounts (such as Sears) and one bank reference are required. Your suppliers will also want the same of your lender. This lender reference is the key to your getting credit with them. Obviously, if your credit is strong enough for the lender, it's strong enough for the suppliers. Often, just the lender reference is enough.

The suppliers you most likely will be using are:

- Sand and gravel company.
- Brick company. For face (decorative) brick.
- Concrete block and brick company. For mortar mix and the building blocks for foundations.
- Concrete supply company. For concrete for basements, garages, footings and driveways.
- Lumber company. For framing lumber, nails, windows, doors, roofing, siding, paneling, and interior and exterior trim.
- Floor covering company. For vinyl flooring, carpet, parquet flooring, and Formica counter tops.

- Lighting fixtures supply. For light fixtures, bathroom fans, exhaust fans. The firm will do a complete count of your lighting needs and help you stay within a dollar figure set during estimation.
- Paint store. For wallpaper. Paint is usually purchased by your painter and included in his price per square foot.
- Appliance store. Some lumber companies and lighting centers carry appliances.
- Insulation company. For your insulation needs.
- Tile company. For ceramic tile, marble, any decorative stonework.
- Drywall company. For Sheetrock. Some lumber companies also sell Sheetrock.
- Any other specialty type suppliers that you may need for certain items not carried by one of the above, such as cabinet shops. Most lumber companies sell cabinets.

An account at a plumbing supply company is not necessary because the plumber sub buys all these items. You may visit to make your selections.

Some of the suppliers listed may carry most of your needs. In some small towns I've seen lumber yards that carry everything from bricks to wallpaper. So long as their prices are competitive, that's fine.

This is my recommendation: when you find a lumber company, give your plans to the sales force and let them give you a list of all the items the company can provide and a complete cost list. I have never found a company that wouldn't do this. Some will do a complete takeoff of all the lumber and other materials you will need. I have it done by a lumber company when I am pressed for time.

You will get a list of the number of studs, floor joists, windows, and rafters. It will be fairly accurate, so don't be afraid to use it. Remember, the company is eager to sell to you and this is part of its service. As the housing market gets slower, the company will become more eager. Company personnel can also explain different products to you and show you new products and ideas. Let all your suppliers aid you in take-offs and different product ideas.

Delivery of Materials

Most companies will also assist you in, or be responsible for, delivery, making certain the materials are delivered on time, but not days in advance so they may be stolen. This delivery system has worked particularly well for me when I have used one supplier for all framing and as much of the rest of the supplies as I could. In such a case I have permitted my carpenter to order needed materials from that supplier alone. The supplier has aided me in keeping an eye on the additional materials ordered.

Buying at Builders' Cost

"Ask and ye shall receive." If you don't, you won't. Be sure to ask if the quoted price is the "builders' price." Tell the supplier that you are a builder, because you are. If you want to add "company" to your name (that is, John Smith Company), fine. If it makes you feel better, do it. It costs nothing and it changes nothing legally (the credit will still be in your name). If you want to incorporate (form a corporation), talk to your attorney. I don't think you'll find it is necessary.

Paying Your Suppliers

Some suppliers offer at least thirty-day terms so that you pay the stated amount within that time without penalty. They also offer 2 percent discounts if your bill is paid by the tenth of the month following purchase. If you buy supplies on June 20, you must pay by July 10 to receive the discount. If you buy on June 1, you would still have until July 10 to pay and get the discount. I try not to order too much toward the end of the month and wait as much as a week, if I can, to get an extra thirty days to pay. Not all suppliers are as generous as this, and their terms will vary. Check with your suppliers on what terms will be offered you.

By getting favorable terms, and extra time in which to pay, you can be sure that your construction draw will cover the bill. You will use construction draws to pay for all your supplies. You should never have to use any of your own money.

Construction draws are based on labor completed and materials in place, not stored or sitting on the job site. For example, if you order brick for the whole house (a *brick veneer* home) at the start of the job, you'll have to pay for the brick that's going to be used in the veneering before the lender advances the dollars for veneering. Instead, you should order the brick in two stages, first for the foundation, then for veneer when you're ready.

Bookkeeping

Bookkeeping is very simple when you are building only one house. Open a separate checking account to handle the construction of the house and pay all bills, no matter how small, by check. Code each check stub to one of the appropriate categories on the estimate sheet and record each amount on a sheet of paper until you have totalled all of the money spent on that category. Then enter the total amount in the actual cost column of the estimate sheet and compare the actual cost to the estimated cost.

Chapter 8

Building the House

Look at how much you have accomplished. You have your building lot, you've arranged your loan, and you have a house plan you can live with, and finance. You've completed the vast amount of paperwork that includes any necessary permits and insurance policies.

You have located most of your subs, and have contracts with them. You've visited various supply houses, and worked out your accounts for bricks, concrete, and lumber.

Congratulations. You've reached the day you thought might never come. You're ready to start building.

What is the proper sequence of steps in building the house, and how long will each take? Let's make a list of them.

1. Staking the lot and house: 1–3 hours.
2. Clearing and excavation: 1–3 days.
3. Order utilities, temporary electric service, and a portable toilet: 1 hour.
4. Footings (steps 3 and 4 can be reversed). First inspection must be made *before* pouring: 1 day.
5. Foundation and soil treatment, then foundation survey: 1 week.
6. Rough-ins for plumbing, if on a slab, and inspection: 2–4 days.
7. Slabs, basement, and garage: 1–2 days.
8. Framing and drying-in: 1–3 weeks.
9. Exterior siding, trim, veneers: 1–3 weeks.

10. Chimney and roof shingles: 2 days–1 week.
11. Rough-ins, can be done while steps 9 and 10 are in progress: 1–2 weeks.
12. Insulation: 3 days.
13. Hardwood flooring and underlayment: 3 days–1 week.
14. Drywall: 2 weeks.
15. Prime walls and "point up": 2 days.
16. Interior trim and cabinets: 1–2 weeks.
17. Painting: 2–3 weeks.
18. Other trims, such as Formica, ceramic tile, vinyl floors: 1 day–1 week.
19. Trim out and finish plumbing, mechanical, and electrical and hook up utilities: 1–2 weeks.
20. Clean up: 2–3 days.
21. Carpet and/or hardwood floor finish: 3 days–1 week.
22. Driveway (if concrete, can be poured anytime after step 14): 1–3 days.
23. Landscaping: 1–3 days.
24. Final inspections, surveys, and closing of construction loan and interim loan: 1–3 days.
25. Enjoy: a lifetime.

(Note that steps 2 and 4 can be done by one sub.)

The Steps Explained

Let's take a closer look at those steps, and clear up any details you should know about them.

1. Staking the Lot and House (1–3 hours)

Since this will be, most likely, the first house you have built, I recommend that you have a registered surveyor/engineer do both. The least expensive one should be the one who surveyed the lot for purchase. Surveying is an important function as houses have been placed in violation of certain setbacks or restrictions, and if a surveyor does this, he's responsible. Builders have placed houses straddling property lines, and have had to tear the foundations out and start over. Best to be safe.

The stakes for the lot may have been moved or torn down since you purchased it. A surveyor will check this. Cost for restaking a lot and staking a house should be $100 to $200 depending upon the complexity of the house and the terrain. You, of course, will want to be

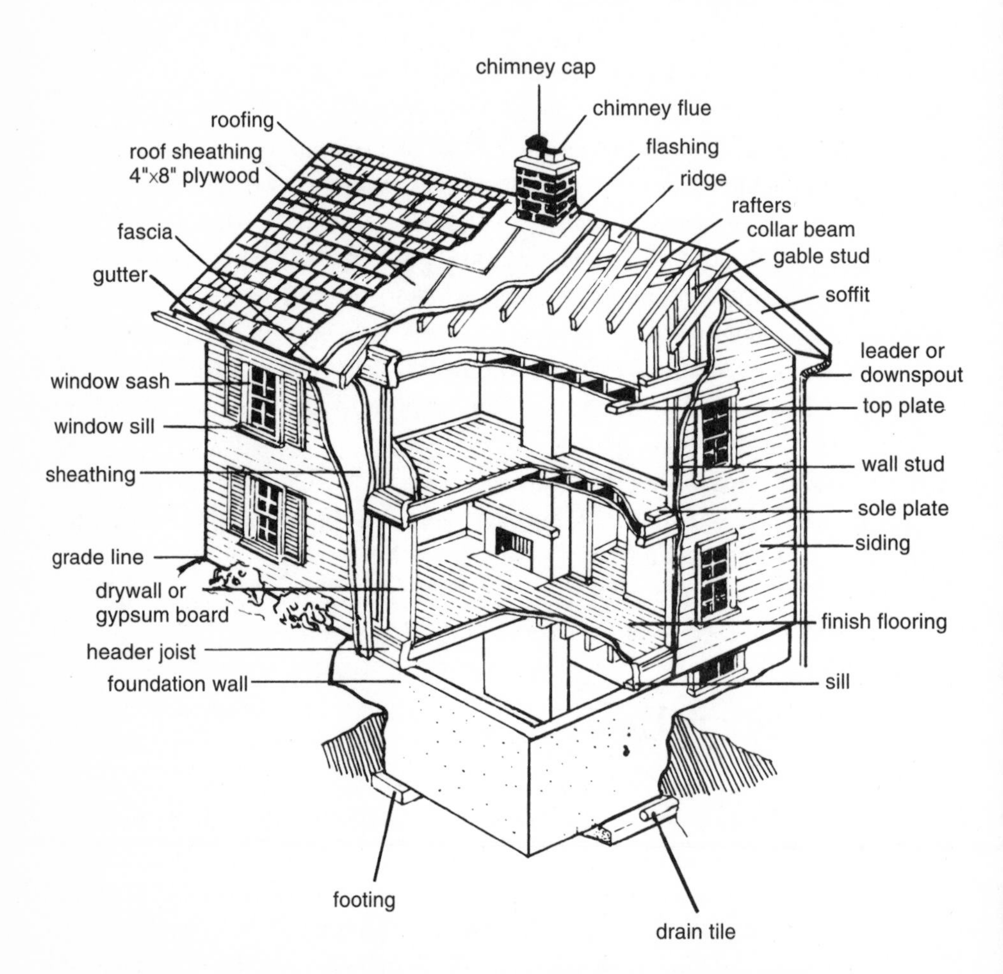

You can be a successful contractor of your own house without knowing a soffit from a fascia, but it helps, in dealing with the subcontractors you hire, if you can speak their language. You'll find in this illustration a sample of the common terms you'll hear used by your carpenters. For other terms of the trade, and some you need to know when dealing with your home financing, see the Glossary on pages 91–94.

present for the staking or placing of the house to be certain it faces the direction that you want. You could meet with the surveyor early in the morning, at noon, or in the late afternoon, or even on a Saturday for this. Or you could put some stakes in the ground over a weekend, and thus indicate the approximate location and direction you desire. Then let the surveyor do it accurately and you can inspect it when it is done.

When the lot is cleared and the basement — if you have one — is dug, you may want the surveyor back to restake the house. On his first visit he usually will put in offset stakes. Offset stakes are the original four or two corners offset as much as twenty-five feet to the side so they won't be disturbed during clearing and excavation. Then it's a quick job for him to find the exact locations on his second visit.

Usually your house should face in the same direction as other houses on your side of the street. It doesn't have to, especially if your house is on a large lot, unless the building code demands this.

If your lot slopes more than three or four feet, you may need a *topographical plat.* Shop for the best price. You need it so that you can fit the house to the slope, and be certain water will go around the house as it drains off the grade.

Water is, has been, and always will be the biggest problem that interferes with the construction of a building. It is nature's strongest force. It has shaped this planet. It will go where it can. It will take the shortest route. It is strong, stronger than anything man can make. Gutters help get rid of water. They don't control it. Keep it in mind.

The house should be positioned first on a map of your lot, then that position should be staked on the site by you and your carpenter, surveyor, or footing contractor. The surveyor is best. An architect or designer can position the house on the lot on paper, considering water as well as solar, wind, light, and the location of other houses. You should contribute your ideas to the person positioning the house. A few things to consider:

- Light. A north-south facing house will be darker than an east-west facing house.
- Water flow.
- Other houses on the street.

- The street it faces. Should the house be parallel with the street? What if the street curves? What about a corner lot?
- Privacy, on the front, back, and sides.
- The desire to use solar energy. Most solar homes have as much surface as possible facing south, with roof solar collectors or broad expanses of glass on that side, and a minimum of doors and windows on the north.
- Minimum setback and side-yard requirements.

I don't want to frighten you with so many considerations. I have built many houses for my family, and I still have to think through these decisions. Positioning the house is a very personal decision, one you should make when you buy a lot. If you are stumped and you can't decide for yourself, and don't have an architect or designer involved with your plans, spend the money and get the advice of an architect for that particular decision.

2. Clearing and Excavation (1–3 days)

Clearing the lot is what the title implies, clearing trees, brush, rocks, roots, and debris from where the house will sit, and usually ten feet around the site, thus allowing space for tractors, fork lifts, and trucks working at the site. Obviously, the more area to be cleared, the more it will cost. Big tree removal is time-consuming and expensive. Rocks may have to be blasted. If you want unwanted trees to be cut into firewood, the crew you hire will charge dearly for doing the job. Best to have the good, manageable-sized logs cut down to ten to twelve feet in length and piled at the side of the lot for you to cut up at your leisure.

Your best source for finding a sub for clearing and excavating is by word of mouth or in the Yellow Pages.

Get a *contract price* for this work. It may cost a little more, but you will be assured of not having your first cost overrun. If a basement is to be dug, your sub must have and know how to use a *transit.* You may also want your surveyor/engineer to oversee this digging to make certain it is the proper depth. Some or all of the dirt removed from the basement site might be put out of the way for later backfilling and landscaping. Topsoil should be separated, to be spread later for the lawn or garden.

Your contract price should include hauling all trash, such as stumps, branches, and rocks, to a suitable landfill. I don't advise burying trash on your lot as these stump holes tend to form an unsightly depression as the material in them settles or rots. Some areas ban stump holes. Of course, if the distance to a suitable landfill is prohibitively expensive and you have enough land, a stump hole may be your best choice.

You may want to put one or more loads of stone on your driveway so that supply trucks can drive in during wet weather. The best stone for this is unwashed crushed stone. It has all the powdery substance created in crushing the stone, and this will harden after getting wet. You also may need to put drainpipes in at the roadside if they are required by either ordinance or common sense. They allow roadside water to flow under the driveway and prevent water from washing the stone away.

3. Utilities Hookup (1 hour)

When you purchased your lot, you were told (be sure you were) what utilities were available and how much they would cost. Now it's time to make plans for a couple of months down the road with a few phone calls and/or a visit to each utility. You should pay all fees and complete any necessary forms. Arrange for temporary electric service for your subs. Usually your electrician is responsible for installing the temporary electrical panel box and having it inspected, but *you* will have to apply for the service from the utility. This usually can be done over the phone.

Wells and *septic systems*, if used, can be installed now, and it is best to get this work done at this time. County or city health inspectors may be required by code to determine the location of these. Tell them your plans for such things as gardens or driveways, or which trees you hope to save, to guide them in their decisions.

If no temporary source of water is available, such as a house next door, you will have to have the well dug and temporarily wired for your brick masons who will be needed shortly, or they will have to truck in their own water.

I also recommend, and some locales require, a portable toilet on the job site. Sources for renting these can be found in the Yellow Pages under Toilet–portable.

4. Footings (1 day)

The footing is the base of a structure. It is a mass of concrete supporting the foundation of the house. It can be poured into wooden forms or in trenches. It must be below the *frost line*, or it will heave when the ground thaws and freezes. In the northern states and higher elevations of any area, it may be four or more feet below grade level. This is one reason there are more basements in a northern climate. If you have to be several feet below the grade for your footing, and thus need several feet of foundation to get back up to grade level, only excavation and a concrete slab are needed for a basement. Local codes will clearly state the requirements for footings in your area. The subcontractor you choose should know the code.

I have a footing sub who stakes, clears, excavates, digs, and pours footings. With a little effort, you can find the same. For your first house, I recommend that you do. The cost will be about the same. The footing is probably the most important part of the house. If it settles or moves, so will your house. If it is not done according to the dimensions of your plans, you will either have to change the plans to accommodate the footing, or do the footing over. I recommend the former if the situation should arise, unless the deviation is too severe to live with.

After your foundation walls are up, you should put in a footing drain. Your code may require this. The drain can be connected to a dry well, storm sewer, or any other approved means of getting rid of the water. In some places, it can simply drain into your yard.

As a rule, footings are better today than they were 100 years ago. Well-built houses of today will probably last more years than those built long ago. Technology has improved materials such as concrete, and our knowledge of how to use them. I say this to help ease your mind about this important step.

Building inspectors usually check the locations of footings before they are poured, to make certain they are deep enough, and resting on undisturbed earth. Don't complain about this inspection — it could save you thousands of dollars if it meant you were avoiding some future problem.

5. Foundations (1 week)

Foundations can be made of brick, concrete block, or poured concrete. Stone foundations, as a rule, aren't built anymore, as they aren't as strong as the others. Stone can be applied as a veneer just like brick, for aesthetic purposes, and you would be wise to use it only as such. Local codes may prohibit stone as a foundation load-bearing material.

Your masonry contractor needs to be one of your better subs. Next to your carpenter, he is the most important. Your carpenter probably can recommend a good mason and he usually starts his work soon after, if not directly after, the mason is finished. He knows the good ones. He's had to, I'm sure, follow bad ones. If he followed a bad one, he'll remember having to *shim* walls or make an out-of-square foundation work as well as he can. Houses *can* have square corners. Most of mine do. (Some don't.)

The foundation wall for any type house needs to be high enough so that water (our old friend) will be diverted away from the house by the final grade of the soil around the house. It must also be high enough so that the wood finish and framing of the house will be at least eight inches above the finish grade and thus protected from soil moisture.

A crawl space should be at least eighteen inches high so that you can crawl beneath the house annually to inspect for such things as termite damage. The crawl space walls should have screened openings for ventilation.

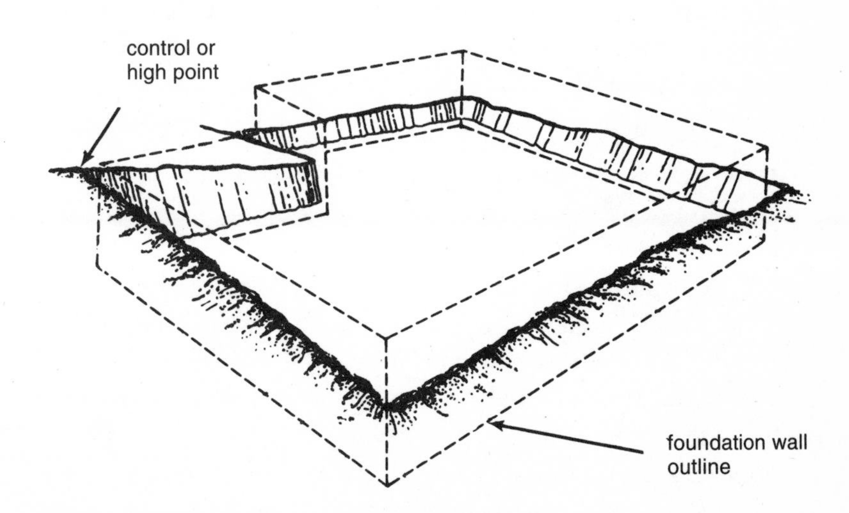

How to establish the depth of excavation.

If you are planning on a full basement, your foundation walls should be high enough so that you have at least seven feet four inches of head space in the finished basement.

If you are in doubt about the foundation height, consult an engineer. Usually you, with the help of the carpenter, mason, excavator, or anyone who can use a transit or a level, can determine the needed height. If the lot is almost flat, the job is simple. It becomes tricky when the lot is steep, or has opposing grades. Experienced contractors make certain the foundation is high enough at the highest point of the outline of the foundation wall, and use that highest point as the control point.

The finished foundation should be *waterproofed* from the footing to the finish gradeline. I recommend hiring a professional waterproofing company for this. Companies are listed in the Yellow Pages under waterproofing. Don't let a laborer do it in his spare time, if he offers to. A professional company will stand behind its work.

Also, depending on your locale, you may need to have the soil treated for insects and pests, particularly termites. Hire a professional. This job is done after the foundation is in, but before any concrete is poured for either the basement or the garage. The cost is small.

It may come as a surprise to many first-time home builders that the foundation is poured or formed with concrete blocks, then holes are punched in it for such things as the water supply and the sewage outlet, the pipe is placed through the hole, and the space between the pipe and wall is patched. This is the easiest method to use to get a tight, waterproof fit.

6. Rough-in Plumbing (2–4 days)

If you have a basement with plumbing or if you are building the house on a concrete slab (as opposed to wood floor joists), once the foundation is in and backfilled and tamped (packed down), and the soil treated, your plumber needs to install the sewer line and the water pipes that will be under the concrete. Also, any wiring that will have to be under the concrete needs to be placed in conduit and roughed-in. Most wiring though, can be run through the stud walls and ceiling joists to any given point.

Your soil treatment company may want to wait until the rough-ins are completed before treating the soil so it won't be disturbed by digging in the plumbing lines. Ask about your company's policy.

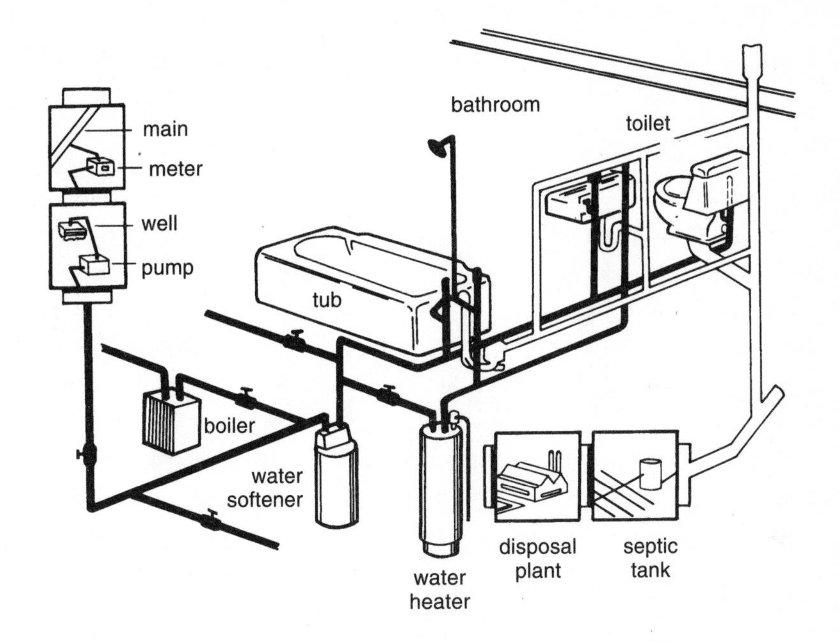

The plumbing system of a home. Note that the supply system pipes are black, the drainage system's are white.

7. Slabs — for Heated Areas (1–2 days)

Many locales require slab perimeter insulation. This is one-inch Styrofoam run around the perimeter four inches high and eighteen inches wide. I recommend it, even if it is not required by codes. I also recommend using a four to six mil thickness vapor barrier of polyurethane (poly) under the concrete to prevent moisture from working up into the concrete. A six by six-inch #10 wire mesh should be placed in the concrete to reinforce it.

The top of the slab should be at least eight inches above the finish grade. Your sub should put down a base for the slab, tamping down gravel or crushed stone to form a layer four to six inches in depth. The poly goes down on this just prior to pouring the concrete. If you cannot cover the entire area with one sheet of poly, any joints of the poly should overlap by four inches and be sealed. The wire mesh is laid on top of the poly. Call for an inspection before pouring concrete if your code requires it.

A good concrete sub will do all of this. I stopped checking my slab pourings when I found a terrific concrete subcontractor.

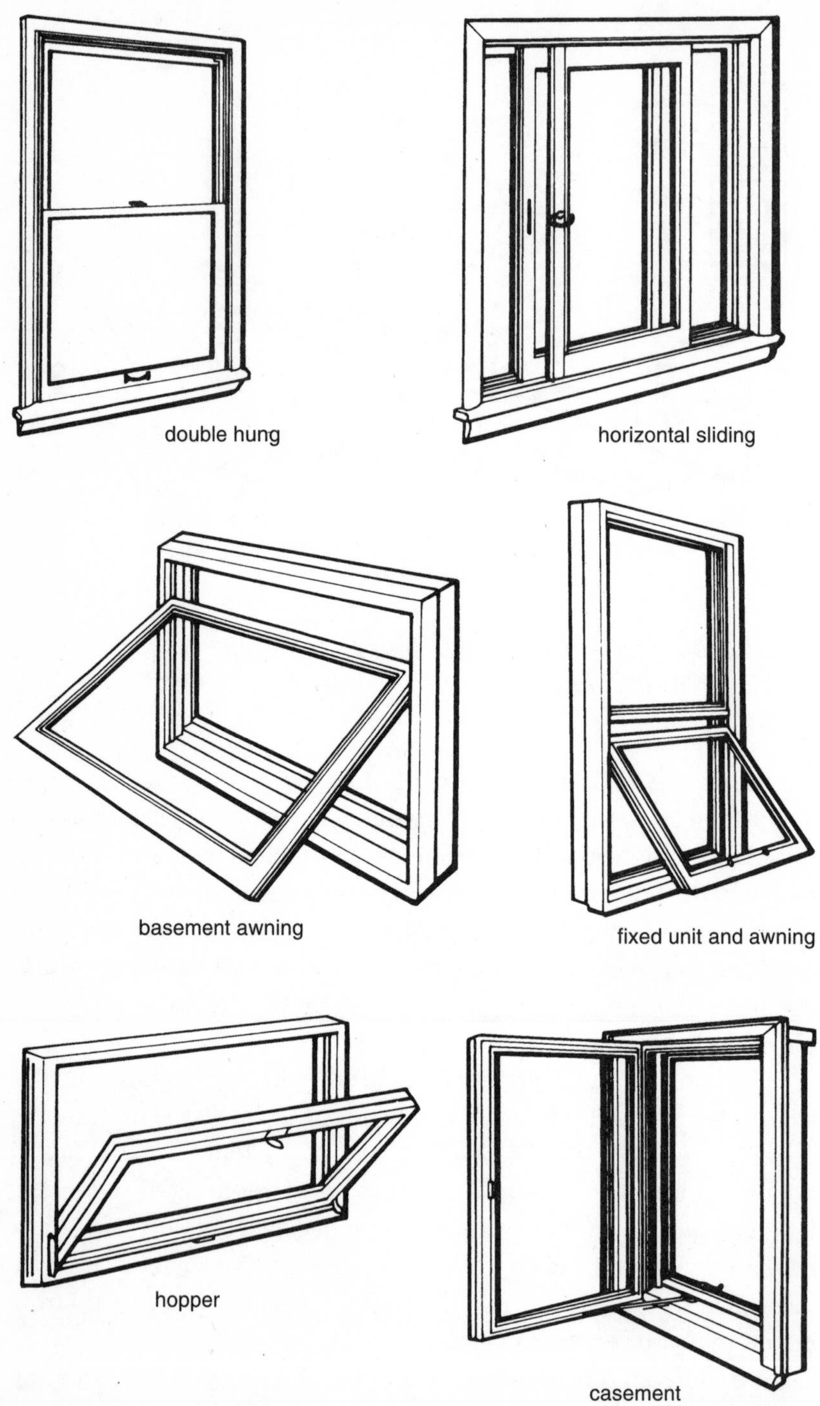

Various types of windows.

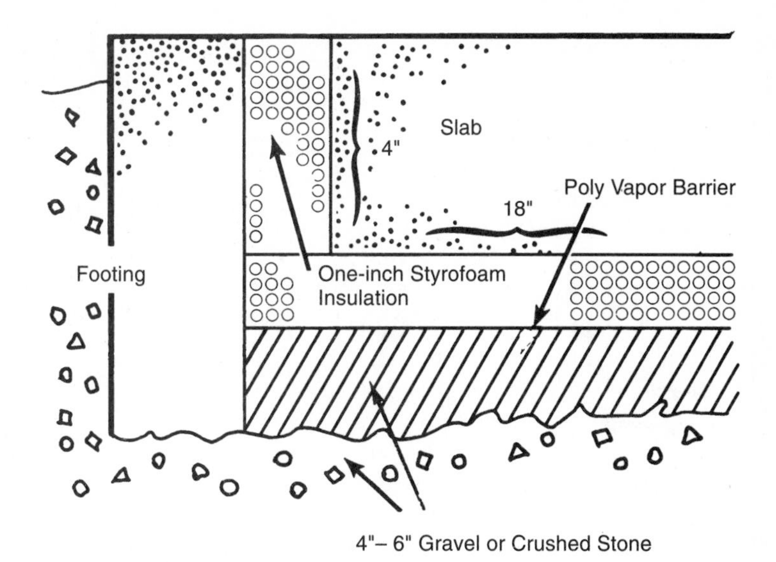

Proper insulation for a concrete slab.

For garage slabs, with proper backfilling and tamping, you don't need the wire mesh, but if you believe the added strength is needed, use it. I recommend an *expansion joint* of fiber board around the perimeter. Garages are subject to extreme temperature changes, and concrete expands and contracts with those changes. The expansion joint permits this expansion without cracking the concrete. Both this slab and the house slab should be at least four inches thick. Be sure your sub thickens all slabs wherever they will be carrying load-bearing posts or walls. Codes differ on the additional thickness required, but it is often the thickness of the footings.

8. Framing (1–3 weeks)

If you have a good carpenter, this is the end of this section on framing. And if that sounds too simple, wait and see. You need only order the lumber, the windows, and the exterior doors, and in two or three weeks you'll have a house (or at least something that looks like a house). I have included here a typical order list for framing and drying-in, which means making the house secure from rain. Rain or snow during framing is not desirable, but it seldom does much damage beyond an occasional warped piece of lumber. But once the house is dried-in, the work inside can progress regardless of the weather if windows and doors are in place.

FRAMING LIST

15	pieces	2 x 6 x 12 treated pine
10	pieces	2 x 4 x 12 treated pine
20	lineal feet	2 x 2
250	lineal feet	1 x 4
18	pieces	2 x 8 x 12
80	pieces	2 x 8 x 14
80	pieces	2 x 8 x 16
55	pieces	2 x 10 x 12
100	pieces	2 x 10 x 14
18	pieces	2 x 10 x 16
18	pieces	2 x 10 x 18
2	pieces	2 x 10 x 20
14	pieces	2 x 6 x 14
45	pieces	2 x 6 x 16
80	pieces	2 x 6 x 20
800	pieces	West Coast studs, 2 x 4 x 93
175	pieces	Half-inch CDX plywood
24	quarts	Plywood glue
10	rolls	#15 felt
60	pieces	Asphalt impregnated sheathing ½ x 4 x 8
300	pounds	16d nails, coated
150	pounds	8d nails, coated
10	pounds	Steel cut masonry 8d
25	pounds	⅞-inch galvanized roofing nails
50	pounds	1¼-inch galvanized roofing nails
10	pounds	16d galvanized finish nails
150	pieces	2 x 4 x 14
50	pieces	2 x 4 x 10

You will want to check with your carpenter about any problems with materials. It is in this stage that I recommend installing a job site telephone. Your carpenter can take it home at night to prevent misuse. If you don't want the expense of a telephone, other arrangements for more frequent communication can be made, such as using a nearby phone, or more frequent visits by you or someone else for you. But since it is impossible to estimate exact needs in materials, some ordering will have to be done. You could give your carpenter permission to order what he needs and then tell the supply house what you have done. The decision is up to you. I do it.

The cross sections of typical framing here are to make you familiar with some of the construction terms.

When the framing is completed, order cabinets, bookcases, and bath vanity cabinets, if there are any. Space for them can now be measured on the job by the salesman.

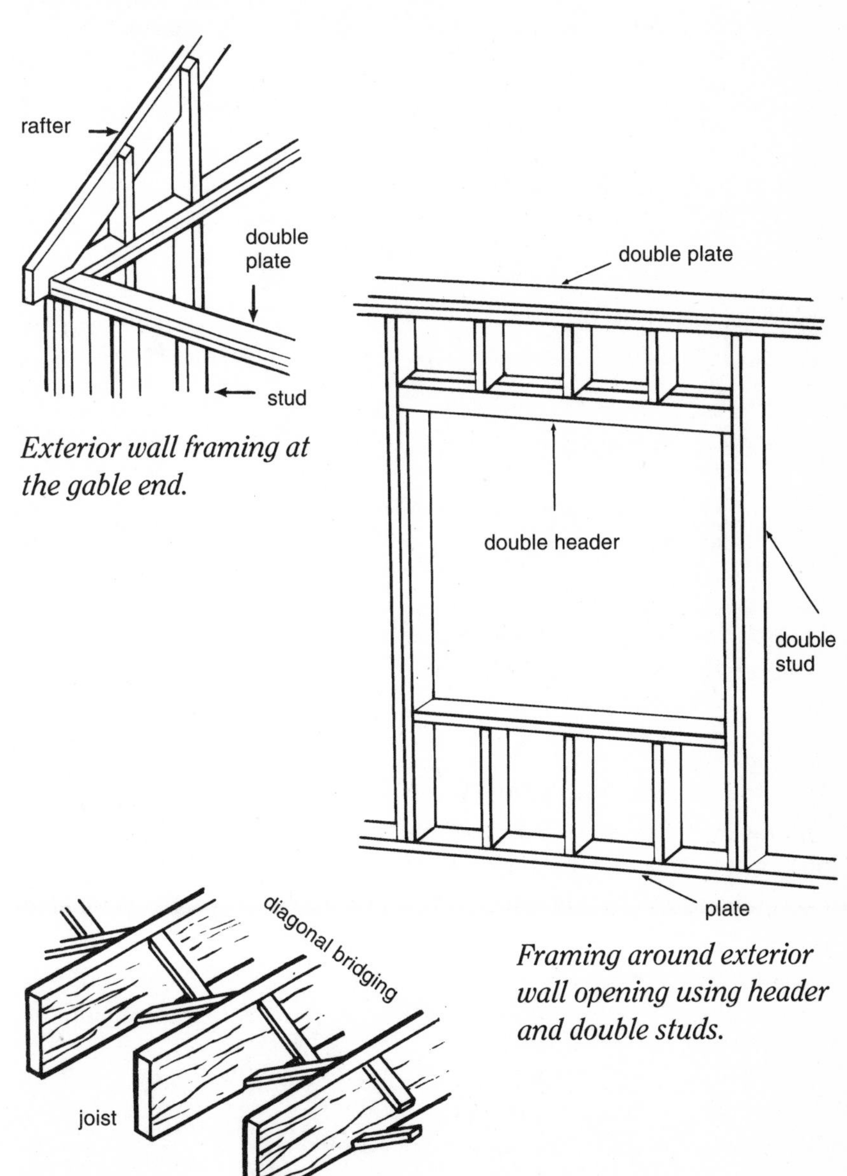

Exterior wall framing at the gable end.

Framing around exterior wall opening using header and double studs.

Diagonal bridging of floor joists.

9. Exterior Siding, Trim, Veneer (1–3 weeks)

This phase of construction is carried on while work progresses inside and should be done before roof shingles are installed. Masonry chimneys are installed after siding or veneer. Veneers such as brick should be installed before final exterior trim (boxing) is added. At completion of this step, you are ready for exterior painting.

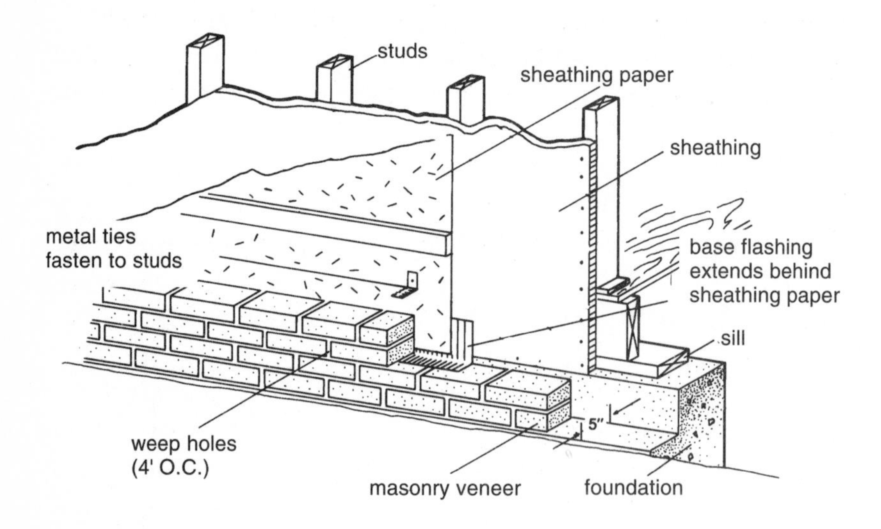

A wood frame wall with masonry veneer. Note shelf on foreground of foundation, to support bricks.

10. Chimneys and Roofing (2 days to 1 week)

Chimneys should be built before the roof is shingled. This will allow placement of sheet metal flashing around the chimney for waterproofing, and will also avoid damage to the shingles. A *prefab fireplace* and *flue* would also be installed at this time. Roofing follows completion of the chimneys.

11. Rough-ins (1–2 weeks)

All electrical, plumbing, heating and air-conditioning, phone prewires, stereo and intercom, burglar alarm systems, should be roughed-in at this time or anytime after step 8 is completed. This does not mean that these units are installed at this time — only the wiring or plumbing for them. Inspections are needed when this step is complete.

12. Insulation (3 days)

Consult with your local utility company on the insulation you need to qualify for their lowest rates. Some locales require an inspection of insulation by both the utility and the building inspection department when it is completed, and before it is covered with drywall or paneling (or plaster, if anyone still wants plaster walls).

13. Hardwood Flooring and Underlayment (3 days–1 week)

I find it easier to install hardwood flooring and carpet or vinyl underlayment before I have the drywall installed. It can be done afterward.

14. Drywall (2 weeks)

Most residential walls are drywalled with ½"×4'×8' gypsum wallboard sheets called drywall or Sheetrock. For baths and other moist areas use a waterproof board, or paint it with an enamel paint, even before wallpapering.

Figure on 3½ or 4 times the square footage of floor area for the total square footage of wallboard to be used. Your drywall sub can give you a price based on a square foot charge. Some will give a bid that includes materials. On your first house you may want to go this route.

15. Prime Walls (2 days)

After the drywall is finished and before the interior trim is applied, I prime all walls and ceilings with a flat white latex paint. My two painters do this with a spray gun and can do in one day a house that measures 3,000 square feet. By priming first, the finish painting time is reduced considerably, thus saving you money. With no trim installed, only windows need to be protected during spraying. If all of your woodwork (trim) is going to be painted, instead of being stained, priming the walls by spraying can be postponed until the painting stage.

When the walls and ceilings are primed, the slight imperfections that sometimes occur in finishing drywall will show up. At this time you can have your drywall subcontractor touch up these places so that the walls and ceilings will be ready for final painting (this is called *pointing up*). Pointing up can also be done just after the interior trim is finished in case the walls get nicked during this process. I have done it *both* times, and at no extra charge.

16. Interior Trim (1–2 weeks)
Doors, moldings, cabinets, and shelves are installed at this time. Cabinets that were ordered after the completion of step 8 should be ready for delivery and installation. The cost of installation should be worked out in advance between you and either your carpenter or supplier (or cabinetmaker if you use shop-built cabinets). Interior trim labor usually includes standard trim and moldings. Any special molding or trim work, or paneling, should be discussed with your carpenter in the planning stages to determine the extra cost of installation. Your millwork supplier, usually the lumber company, can do a complete material take-off for your trim and discuss costs of extra items.

17. Painting (2–3 weeks)
You are ready for final painting inside. Your exterior painting can be delayed until this time, too, to save your painter from hopping back and forth from job to job. Work this out when you first discuss painting with your sub. You don't want to leave the exterior trim unpainted or unstained too long, as it may warp or get moldy.

18. Other Trims (1 day–1 week)
It's time to install Formica, vinyl floors, and ceramic tiles. Wallpaper can be done at this time, or delayed until after you move in. The others can't be delayed, because plumbing cannot be completed until they are.

19. Trim Out (1–2 weeks)
It's time for the plumber to finish his work. This is called trimming out or setting fixtures. Some of the fixtures he installs must be wired, so he needs to finish before the electrician can finish. Also, your heating and air-conditioning must be completed before your electrician finishes his work. Next comes your electrician, to install switches, receptacles (sometimes called devices), light fixtures, and electrical appliances, such as the oven and range. He will also wire the electrical apparatus that has been installed by your plumber and heating and air-conditioning subs.

20. Clean up (2–3 days)
Both inside and outside cleaning can be started. The bulk of outside trash, and an incredible amount of it that comes from inside, can be picked up by truck and hauled away. You can do the inside work. If you don't want to, you can get a professional. Be sure to add the cost to your estimate.

21. Carpet (3 days–1 week)
Hardwood floors should be finished before the carpet is installed. Allow at least three days for hardwood finishing. No other subs should be working in the house while this work is done. They are expensive floors, and it is well worth a few days of "no trespassing" to have them done right. They should be finished before the carpet is installed because of the sanding required before the floors are stained and sealed with polyurethane. The dust from sanding usually is controlled, but some could permeate the rooms with carpeted floors.

Because of this dust, you may want to finish the floors before step 20, cleaning up. I've done it both ways, and found advantages to both.

22. Driveway (1–3 days)
You should keep heavy trucks off newly laid concrete or asphalt drives for a period of time. Try to time it so that you have received all heavy shipments of materials, and hauled off all heavy loads of trash, and don't have any heavy equipment coming to do work. Realize that if you wait until last with the driveway, you'll at least have a moving van on it. Concrete can take a moving van a week after the concrete is poured; not so with asphalt. If you use asphalt, wait until after you move in to pave. Put down a good stone base on the drive before moving in. My asphalt paving contractor insists on this and puts down the stone base and then redoes the stone (or dresses it up) before paving it with asphalt several days after the moving van leaves.

23. Landscaping (1–3 days)
This is an item that, depending on the requirements of your lender, can be put off until after you move in. Such a situation may arise due to weather or scheduling problems or lack of money due to other cost overruns at the end of construction. You might get by with just

grading and seeding, or mulching disturbed wooded areas.

24. Final Inspections, Surveys, Loan Closings (1–3 days)
After the completion of the actual house (not the drive and landscaping), all final inspections from the county or city, for building, electrical, mechanical, and plumbing should be made. Also, the lender will make a final inspection at this time, and may require drives and landscaping to be complete before it will disburse the balance of the construction loan and close it out to the permanent mortgage.

When approval is given, your attorney will coordinate the necessary paperwork and schedule the loan(s) closing. The lender will most likely require a final survey to be sure no additional structures or additions have been placed on the lot in violation of deed restrictions or zoning. Generally, your attorney will order this final survey and any other necessary documents required for closing.

The actual loan closing will only take about fifteen minutes, depending on the efficiency of the attorney and the lender. One document you will want to remember on your own, but you will be reminded of by your attorney or lender, is your insurance policy. It will need to be converted into a homeowners' policy prior to closing. This merely takes a phone call to the agent who issued you your builders' risk (or fire) policy.

25. Enjoy!

Chapter 9

Additions to an Existing Home

If you already have a house, but would like more room and don't want to move, you can add to your present home. This option would also require less money than building a bigger house, since you have a house purchased when houses were cheaper to build and mortgages cost less.

If your mortgage interest rate is low, it is difficult to think of exchanging that rate for higher rates. An addition may be the solution for you. It could be easier than you think. You may already have a head start toward additional space.

There are two basic ways to add to a home. One is to make habitable an existing unfinished area such as a basement, garage, screened porch, breezeway, or attic. This is the least expensive way because you already have a room, foundation, and some or all of the outside walls. The other and more expensive way is to build on to your house. If you consider this, check to see that you have room on your lot without violating required setbacks and that you meet any subdivision or deed restrictions.

Don't add too much value to your home. You don't want to have the most expensive house in the neighborhood. It's a real estate fact of life: the $100,000 house in the $50,000 neighborhood is very difficult to sell.

Whether you finish an existing space or add to your present structure, the procedure will be the same. Start at the beginning of this book and treat the addition as a small home. The only things that will be different from building a house are that you already have the land and the costs will be somewhat different. Don't skip any of the applicable steps I outlined earlier for building a house. You will need an attorney to check restrictions and for the loan, plans, an estimate, and the money to build. Look at the end of Chapter 5 (pages 56–57) before deciding to do your own labor.

How to Estimate Costs

The basic differences in estimating an addition against a complete home will be the elimination of some costs due to the existing structure and the increase of some square foot costs due to the smaller size of the job. Subs may want a few more dollars for their labor costs because they could be earning more for almost the same amount of time.

The general contractor that you won't need would have charged a greater percentage for profit and overhead for the same reason as the subs. This savings will more than offset the increase in labor costs from subs.

You should obtain bids (quotes), as though you were building a house, and proceed from there.

How to Finance an Addition

Financing an addition is far easier than financing a new home. You can simply get a *second mortgage* based on the equity you have in your home. Your equity is the difference between what the house is worth and what you owe on it. Quite often you don't even need to give a reason to borrow against that equity because the lender is well protected if you should default. The house exists, it is not being built. The amount you can borrow varies from lender to lender but if you have been in your house for at least three years it is safe to assume you have built up some equity. As a rule of thumb you can borrow up to one-half of your equity. The second mortgage, which is usually a five- to ten-year loan, is less expensive per month and in total pay back than refinancing the whole house or building a new house of equal dollar value at today's high mortgage rates.

Just as in planning new home construction, your local lender can sit down with you and discuss your needs.

Glossary

Appraised Value. An estimate of the value of property, such as a house.

Breaker. Electrical circuit breaker. The modern version of the old-time fuse.

Bridge Loan. A short-term loan to bridge the time between the purchase of one house and the sale of another.

Bridging. Small pieces of wood or metal used to brace floor joists.

Brick Veneer. Brick used in lieu of siding.

Carpet Underlayment. Either plywood or pressed wood (particle board) placed over the subfloor to make the floor more solid.

Clear Title. A title (proof of ownership) of any property (land, auto, house) that is free of liens, mortgages, judgments, or any other encumbrance.

Color Run. Materials produced using the same batch of dye, such as bricks, carpet, or paint. Subsequent batches may vary in color.

Contract Price. A pre-agreed set price for a service or product.

Course. One row of bricks.

Crawl Space. A shallow space under a house or porch or in an attic.

Drip Cap. A protective molding, usually metal, to divert water over an exterior surface.

Dry-In. A stage in construction at which time the building is protected from rain and snow.

Drywall. Sheetrock.

Equity. The difference between what you owe on something, and what it is worth.

Expansion Joint. A joint used to separate blocks or units of concrete to prevent cracking due to expansion as a result of temperature changes.

FASCIA BOARD. The flat vertical board at the edge of the roof.

FLUE. A channel for the passage of hot gases and smoke.

FOOTING. A mass of concrete below the frost line supporting the foundation or piers of a house.

FRAME CONSTRUCTION. Using wood, as opposed to brick, block, concrete, or steel, for building walls, ceilings, and roofs.

FRAMING. The construction of the skeleton of a house, to include walls, floors, ceiling, and roof.

FROST LINE. The depth in the earth at which the warmth of the earth prevents freezing or the formation of frost.

HEAT PUMP. An apparatus for heating and cooling a building. In the heating phase, it draws heat from the outside air and transfers it to the inside of a house or other building.

INTERIM FINANCING. A short-term loan. It is generally converted to a long-term loan at a later date.

JOIST. One of the horizontal small timbers ranged parallel and used to support a floor or ceiling.

LOAD-BEARING CAPABILITY. The amount of weight a particular substance can withstand without breaking or bending beyond its design. Used to describe soil, steel, and wood.

LOAN CLOSING. Signing the legal documents with a lender in order to receive a loan.

LOT SUBORDINATION. A process of buying land by which the owner will take a note in lieu of payment and legally take an interest in that land secondary to a party with primary interest such as a savings and loan.

MANAGER'S CONTRACT. A contract with a general contractor by which he agrees to act as a manager to construct your house. Under such a contract, you remain the primary general contractor.

MOLDING. The interior trim in a house.

MORTAR. A mixture of cement, sand, and water used between bricks or concrete blocks to hold them together.

NOTE. A written acknowledgement of a debt, such as a promissory note.

PANEL. Electrical panel. A metal box containing circuit breakers.

PIER. A vertical structural support. Usually a masonry or metal column used to support the house, porch, or deck.

POLYURETHANE. Plastic film, used in home construction to provide a moisture barrier.

PREFAB FIREPLACE. A fireplace that is made of metal as opposed to solid masonry. It usually has a metal flue.

PREWIRE. Wiring for various items such as lighting, telephones, intercoms, burglar alarms, and installed before drywall or paneling is applied to walls.

QUOTE. A guaranteed price in advance.

RECORDING FEE. The fee charged to record legal documents in a place of permanent records, such as a county courthouse.

REINFORCING ROD. A steel rod placed in concrete to increase the strength of the concrete.

ROOF PITCH. The slope of the roof.

ROUGH-IN. The installation of wiring, plumbing, or heat ducts in the walls, floors, or ceilings before those walls, floors, or ceilings are covered with drywall, plaster, or paneling.

SAW SERVICE. Temporary electrical service used during construction.

SECOND MORTGAGE. The pledging of property to a lender as security for repayment, but using property that has already been pledged for a loan.

SEPTIC SYSTEM. A means of disposing of sewage in the ground.

SHIM. A thin wedge of wood or metal used to fill in a space.

SOFFIT. The underside of a roof overhang.

SOFFIT VENT. Air vent in the soffit to allow air circulation under the roofing and between the roof rafters. Prevents heat build-up and rotting of the wood members in the roof.

STAKING. Placing stakes in the ground prior to building to show the location of the corners of the house.

STUD. One of the uprights in the framing of a wall.

Take-Off. The compilation of a list of materials used for a particular phase of construction, such as the number of bricks or the number and sizes of windows. Also called a schedule of materials.

Tamped. Packed down. Tamping soil prevents later settling, such as underneath concrete.

Tax Stamp. A stamp affixed to a legal document to indicate that a tax has been paid.

Test Boring. Sample of the soil upon which a structure is to be built to determine what weight the soil is capable of carrying.

Title Insurance. An insurance policy that protects the owner of real estate from any loss due to defects in his title.

Topographical Plat. A drawing showing the surface features of property.

Transit. A surveying instrument used to measure horizontal angles, levelness, and vertical depth.

Unsecured Loan. A loan in which no material possessions are pledged as security for repayment.

Waterproofing. Making a foundation impervious to water.

Appendix

The following legal instruments are published as examples. Because of varying state laws, these should not be used by you unless such use is approved by your attorney.

Manager's Construction Contract

1. General

This contract dated ________ is between ____________________ (Owner) and ______________________(Manager), and provides for supervision of construction by Manager of a residence to be built on Owner's Property at ________________________________, and described as ____. The project is described on drawings dated ____ and specifications dated ________, which documents are a part hereof.

2. Schedule

The project is to start as near as possible to ______________, with anticipated completion _________months from starting date.

3. Contract Fee and Payment

3A. Owner agrees to pay Manager a minimum fee of ______ ($ ______________) for the work performed under this contract,

a.	Down payment — due prior to start of work	$ ______
b.	Framed	$ ______
c.	Roof on	$ ______
d.	Ready for drywall	$ ______
e.	Trimmed out	$ ______
f.	Final	$ ______

3B. Payments billed by Manager are due in full within ten (10) days of bill mailing date.

3C. Final payment to Manager is due in full upon completion of residence; however, Manager may bill upon "Substantial completion" (see Paragraph 11 for the definition of terms) the amount of the final payment less 10 percent of the value of work yet outstanding. In such a case, the amount of the fee withheld will be billed upon completion.

4. General Intent of Contract

It is intended that the Owner be in effect his own "General Contractor" and that the Manager provide the Owner with expert guidance and advice, and supervision and coordination of trades and material delivery. It is agreed that Manager acts in a professional capacity and simply as agent for Owner, and that as such he shall not assume or incur any pecuniary responsibility to contractor, subcontractors, laborers, or material suppliers. Owner will contract directly with subcontractors, obtain from them their certificates of insurance and releases of liens. Similarly, Owner will open his own accounts with material suppliers and be billed and pay directly for materials supplied. Owner shall insure that insurance is provided to protect all parties of interest. Owner shall pay all expenses incurred incompleting the project, except Manager's overhead as specifically exempted in Paragraph 9. In fulfilling his responsibilities to Owner, Manager shall perform at all times in a manner intended to be beneficial to the interests of the Owner.

5. Responsibilities of Manager

General

Manager shall have full responsibility for coordination of trades, ordering materials and scheduling of work, correction of errors and conflicts, if any, in the work, materials, or plans, compliance with applicable codes, judgment as to the adequacy of trades' work to meet standards specified, together with any other function that might reasonably be expected in order to provide Owner with a single source of responsibility for supervision and coordination of work.

Specific Responsibilities

1. Submit to Owner in a timely manner a list of subcontractors and suppliers Manager believes competent to perform the work at competitive prices. Owner may use such recommendations or not at his option.
2. Submit to Owner a list of items requiring Owner's selection, with schedule dates for selection indicated, and recommended sources indicated.
3. Obtain in Owner's name(s) all permits required by governmental authorities.
4. Arrange for all required surveys and site engineering work.
5. Arrange for all the installation of temporary services.

6. Arrange for and supervise clearing, disposal of stumps and brush, and all excavating and grading work.
7. Develop material lists and order all materials in a timely manner, from sources designated by Owner.
8. Schedule, coordinate, and supervise the work for all subcontractors designated by Owner.
9. Review, when requested by Owner, questionable bills and recommend payment action to Owner.
10. Arrange for common labor for hand digging, grading, and cleanup during construction, and for disposal of construction waste.
11. Supervise the project through completion, as defined in Paragraph 11.

6. Responsibilities of Owner

Owner agrees to:

1. Arrange all financing needed for project, so that sufficient funds exist to pay all bills within ten (10) days of their presentation.
2. Select subcontractors and suppliers in a timely manner so as not to delay the work. Establish charge accounts and execute contracts with same, as appropriate, and inform Manager of accounts opened and of Manager's authority in using said accounts.
3. Select items requiring Owner selection, and inform Manager of selections and sources on or before date shown on selection list.
4. Inform Manager promptly of any changes desired or other matters affecting schedule so that adjustments can be incorporated in the schedule.
5. Appoint an agent to pay for work and make decisions in Owner's behalf in cases where Owner is unavailable to do so.
6. Assume complete responsibility for any theft and vandalism of Owner's property occurring on the job. Authorize replacement/repairs required in a timely manner.
7. Provide a surety bond for his lender if required.
8. Obtain release of liens documentation as required by Owner's lender.

9. Provide insurance coverage as listed in Paragraph 12.
10. Pay promptly for all work done, materials used, and other services and fees generated in the execution of the project, except as specifically exempted in Paragraph 9.

7. Exclusions

The following items shown on the drawings and/or specifications are NOT included in this contract, insofar as Manager supervision responsibilities are concerned:
(List below)

8. Extras/Changes

Manager's fee is based on supervising the project as defined in the drawings and specifications. Should additional supervisory work be required because of Extras or Changes occasioned by Owner, unforeseen site conditions, or governmental authorities, Manager will be paid an additional fee of 15 percent of cost of such work. Since the basic contract fee is a *minimum fee*, no downward adjustment will be made if the scope of work is reduced, unless contract is cancelled in accordance with Paragraphs 14 or 15.

9. Manager's Facilities

Manager will furnish his own transportation and office facilities for Manager's use in supervising the project at no expense to Owner. Manager shall provide general liability and workmen's compensation insurance coverage for Manager's direct employees only at no cost to Owner.

10. Use of Manager's Accounts

Manager may have certain "trade" accounts not available to Owner which Owner may find it to his advantage to utilize. If Manager is billed and pays such accounts from Manager's resources, Owner will reimburse Manager within ten (10) days of receipt of Manager's bill at cost plus 8 percent of suchmaterials/services.

11. Project Completion

a. The project shall be deemed completed when all the terms of this contract have been fulfilled, and a Residential Use Permit has been issued.

b. The project shall be deemed "substantially complete" when a Residential Use Permit has been issued, and less than Five Hundred Dollars ($500) of work remains to be done.

12. Insurance

Owner shall insure that workmen's compensation and general liability insurance are provided to protect all parties of interest and shall hold Manager harmless from all claims by subcontractors, suppliers and their personnel, and for personnel arranged for by Manager in Owner's behalf, if any.

Owner shall maintain fire and extended coverage insurance sufficient to provide 100 percent coverage of project value at all stages of construction, and Manager shall be named in the policy to insure his interest in the project.

Should Owner or Manager determine that certain subcontractors, laborers, or suppliers are not adequately covered by general liability or workmen's compensation insurance to protect Owner's and/or Manager's interests, Manager may, as agent of Owner, cover said personnel on Manager's policies, and Owner shall reimburse Manager for the premium at cost plus 10 percent.

13. Manager's Right to Terminate Contract

Should the work be stopped by any public authority for a period of thirty (30) days or more through no fault of the Manager, or should work be stopped through act or neglect of Owner for ten (10) days or more, or should Owner fail to pay Manager any payment due within ten (10) days written notice to Owner, Manager may stop work and/or terminate this contract and recover from Owner payment for all work completed as a proration of the total contract sum, plus 25 percent of the fee remaining to be paid if the contract were completed as liquidated damages.

14. Owner's Right to Terminate Contract

Should the work be stopped or wrongly prosecuted through act or neglect of Manager for ten (10) days or more, Owner may so notify Manager in writing. If work is not properly resumed within ten (10) days of such notice, Owner may terminate this contract. Upon termination, entire balance then due Manager for that percentage of work then completed, as a proration of the total contract sum, shall be due and payable and all further liabilities of Manager under this contract shall cease. Balance due to Manager shall take into account any additional cost to Owner to complete the house occasioned by Manager.

15. Manager/Owner's Liability for Collection Expenses

Should Manager or Owner respectively be required to collect funds rightfully due him through legal proceedings, Manager or Owner respectively agrees to pay all costs and reasonable Attorney's fees.

16. Warranties and Service

Manager warrants that he will supervise the construction in accordance with the terms of this contract. No other warranty by Manager is implied or exists.

Subcontractors normally warrant their work for one year, and some manufacturers supply yearly warranties on certain of their equipment; such warranties shall run to the Owner and the enforcement of these warranties is in all cases the responsibility of the Owner and not the Manager.

(Manager)__________________________(seal) Date:

(Owner)__________________________(seal) Date:

(Owner)__________________________(seal) Date:

Contract to Build House

(Cost Plus Fee)

Contractor: ______________________________
Owner: _________________________________ Date: ____________

Owner is or shall become fee simple owner of a tract or parcel of land known or described as: ____________________________ .

Contractor hereby agrees to construct a residence on the above described lot according to the plans and specifications identified as: Exhibit A — plans and specifications drawn _______________ by ____ .

Owner shall pay Contractor for the construction of said house cost of construction and a fee of __________. Cost is estimated in Exhibit B. Each item in Exhibit B is an estimate and is not to be construed as an exact cost.

Owner shall secure/has secured financing for the construction of said house in the amount of cost plus fee, which shall be disbursed by a savings and loan or bank from time to time as construction progresses, subject to a holdback of no more than 10 percent. Owner hereby authorizes Contractor to submit a request for draws in the name of Owner under such loan up to the percentage completion of construction and to accept said draws in partial payment hereof. In addition, it is understood that the Contractor's fee shall be paid in installments by the savings and loan or bank at the time of and as a part of each construction draw as a percentage of completion, so that the entire fee shall be paid at or before the final construction draw.

Contractor shall commence construction as soon as feasible after closing of the construction loan and shall pursue work to a scheduled completion on or before seven months from commencement, except if such completion shall be delayed by unusually unfavorable weather, strikes, natural disasters, unavailability of labor or materials, or changes in the plans or specifications.

Contractor shall build the residence in substantial compliance with the plans and specifications and in a good and workmanlike manner, and shall meet all building codes. Contractor shall not be responsible for failure of materials or equipment not Contractor's fault. Except as herein set out, Contractor shall make no representations or warranties with respect to the work to be done hereunder.

Owner shall not occupy the residence and Contractor shall hold the keys until all work has been completed and all monies due Contractor hereunder shall have been paid.

Owner shall not make changes to the plans or specifications until such changes shall be evidenced in writing, the costs, if any, of such changes shall be set out, and the construction lender and Contractor shall have approved such changes. Any additional costs thereof shall be paid in advance, or payment guaranteed in advance of the work being accomplished.

Contractor shall not be obligated to continue work hereunder in the event Owner shall breach any term or condition hereof, or if for any reason the construction lender shall cease making advances under the construction loan upon proper request thereof.

Any additional or special stipulations attached hereto and signed by the parties shall be and are made a part hereof.

Owner: ______________________________(seal)

______________________________(seal)

Contractor: __________________________(seal)

Contract to Build House

(Contract Bid)

Contractor: ______________________________
Owner: ________________________________ Date: ____________

Owner is or shall become fee simple owner of a tract or parcel of land known or described as: ____________________________.

Contractor hereby agrees to construct a residence on the above described lot according to the plans drawn by ___________________ , and the specifications herein attached.

Owner shall pay Contractor for the construction of said house $ ____________.

Prior to commencement hereunder, owner shall secure financing for the construction of said house in the amount of $ ______________ , which loan shall be disbursed from time to time as construction progresses, subject to a holdback of no more than 10 percent. Owner hereby authorizes Contractor to submit a request for draws in the name of the Owner from the savings and loan, or similar institution, up to the percentage completion of construction and to accept said draws in partial payment thereof.

Contractor shall commence construction as soon as feasible after closing and shall pursue work to a scheduled completion on or before seven months from commencement, except if such completion shall be delayed by unusually unfavorable weather, strikes, natural disasters, unavailability of labor or materials, or changes in the plans and specifications.

Contractor shall build the residence in substantial compliance with the plans and specifications and in a good and workmanlike manner, and shall meet all building code requirements. Contractor shall not be responsible for failure of materials or equipment not Contractor's fault. Except as herein set out, Contractor shall make no representations or warranties with respect to the work to be done hereunder.

Owner shall not occupy the residence and Contractor shall hold the keys until all work has been completed and all monies due Contractor hereunder shall have been paid.

Owner shall not make any changes to the plans and specifications until such changes shall be evidenced in writing, the costs, if any, of such changes shall be set out, and any additional costs thereof shall be paid in advance of the work being accomplished.

Contractor shall not be obligated to continue work hereunder in the event Owner shall breach any term or condition hereof, or if for any reason the construction draws shall cease to be advanced upon proper request thereof.

Any additional or special stipulations attached hereto and signed by the parties shall be and are made a part hereof.

Contractor: ________________________(seal)

Owner: ________________________(seal)

________________________(seal)

Description of Materials

☐ Proposed Construction No. ______________________
(To be inserted by FHA or VA)
☐ Under Construction

Property address______________ City__________ State____________

Mortgagor or Sponsor____
(Name) (Address)

Contractor or Builder____
(Name) (Address)

Instructions

1. For additional information on how this form is to be submitted, number of copies, etc., see the instructions applicable to the FHA Application for Mortgage Insurance or VA Request for Determination of Reasonable Value, as the case may be.
2. Describe all materials and equipment to be used, whether or not shown on the drawings, by marking an X in each appropriate check-box and entering the information called for in each space. If space is inadequate, enter "See misc." and describe under item 27 or on an attached sheet. The use of paint containing more than one percent lead by weight is prohibited.
3. Work not specifically described or shown will not be considered unless required, then the minimum acceptable will be assumed. Work exceeding minimum requirements cannot be considered unless specifically described.
4. Include no alternates, "or equal" phrases, or contradictory items. (Consideration of a request for acceptance of substitute materials or equipment is not thereby precluded.)
5. Include signatures required at the end of this form.
6. The construction shall be completed in compliance with the related drawings and specifications, as amended during processing. The specifications include this Description of Materials and the applicable Minimum Property Standards.

1. Excavation:

Bearing soil, type ______________________________

2. Foundations:

Footings: concrete mix __________ ; strength psi Reinforcing __________

Foundation wall: material __________ Reinforcing __________

Interior foundation wall: material __________ Party foundation wall __________

Columns: material and sizes __________ Piers: material and reinforcing __________

Girders: Material and sizes __________ Sills: material __________

Basement entrance areaway __________ Window areaways __________

Waterproofing ______________________________

Footing drains __________

Termite protection __________

Basementless space: ground cover __________ ; insulation __________ ;

foundation vents __________

Special foundations ______________________________

Additional information: ______________________________

3. Chimneys:

Material __________ Prefabricated *(make and size)* __________

Flue lining: material __________ Heater flue size __________

Fireplace flue size __________

Vents (*material and size*): gas or oil heater __________ ;

water heater __________ Additional information: __________

4. Fireplaces:

Type: ☐ solid fuel; ☐ gas-burning; ☐ circulator *(make and size)* __________

Ash dump and clean-out __________ Fireplace: facing __________ ;

lining __________ ; hearth __________ ;

mantel __________ Additional information: __________

5. Exterior Walls:

Wood frame: wood grade, and species __________ ☐ Corner bracing.

Building paper or felt ________ Sheathing ________ ;

thickness ________ ; width ________ ;

☐ solid; ☐ spaced ________ "o.c.; ☐ diagonal: ________

Siding ________ ; grade ________ ;

type ________ ; size ________ ; exposure ________ "; fastening ________

Shingles ________ ; grade ________ ;

type ________ ; size ________ ;

exposure ________ ; fastening ________

Stucco ________ ; thickness ________ ";

Lath ________ weight ________ lbs.

Masonry veneer ________ Sills ________ Lintels ________ Base flashing ________

Masonry: ☐ solid ☐ faced ☐ stuccoed; total wall thickness ________ ";

facing thickness ________ "; facing material ________

Backup material ________ ; thickness ________ "; bonding ________

Door sills ________ Window sills ________ Lintels ________ Base flashing ________

Interior surfaces: dampproofing, ________ coats of ________ ; furring ________

Additional information: ________

Exterior painting: material ________ ; number of coats ________

Gable wall construction: ☐ same as main walls; ☐ other construction

6. Floor Framing:

Joists: wood, grade, and species ________ ; other ________ ;

bridging ________ ; anchors ________

Concrete slab: ☐ basement floor; ☐ first floor: ☐ ground supported;

☐ self-supporting; mix ________ ; thickness ________ ;

reinforcing ________ ; insulation ________ ; membrane ________

Fill under slab: material ________ ; thickness ________ ".

Additional information: ________

7. Subflooring: *(Describe underflooring for special floors under item 21.)*

Material: grade and species ________ ; size ________ ; type ________

Laid: ☐ first floor; ☐ second floor; ☐ attic _____________ sq.ft.; ☐ diagonal; ☐ right angles.

Additional information: ____

8. Finish Flooring: *(Wood only. Describe other finish flooring under item 21.)*

Location	Rooms	Grade	Species	Thickness	Width	Bldg. Paper	Finish

First floor ____

Second floor ____

Attic floor______sq.ft. ____

Additional information: ____

9. Partition Framing:

Studs: wood, grade, and species___________ size and spacing _______ Other ___________

Additional information: ____

10. Ceiling Framing:

Joists: wood, grade, and species _________ Other _________________ Bridging __________

Additional information: ____

11. Roof Framing:

Rafters: wood, grade, and species ____

Roof trusses (see detail): grade and species ____

Additional information: ____

12. Roofing:

Sheathing: wood, grade and species ______ ☐ solid; ☐ spaced_______ "o.c.

Roofing ________________ ; grade ____________ ; size ____________; type ____________

Underlay ________________ ; weight or thickness ; size ____________; fastening ________

Built-up roofing __________ ; number of plies ____ ; surfacing material _________________

Flashing: material ________ ; gauge or weight ____ ; ☐ gravel stops; ☐ snow guards

Additional information ____

13. Gutters and Downspouts:

Gutters: material _________ ; gauge or weight ____ ; size ___________ ; shape ____________

Downspouts: material_________________________ ; gauge or weight _ ; size ______________ ;

shape __________________ ; number __________

Downspouts connected to: ☐ Storm sewer; ☐ sanitary sewer; ☐ dry-well

☐ Splash blocks: material and size ______________________________

Additional information ______________________________

14. Lath and Plaster:

Lath ☐ walls, ☐ ceilings: material ______________________ ; weight or thickness

Plaster: coats ______________ ; finish ______________________

Dry-wall ☐ walls, ☐ ceilings: material _____ ; thickness __________ ; finish __________

Joint treatment ______________________________

15. Decorating: *(Paint, wallpaper, etc.)*

Rooms	Wall Finish Material and Application	Ceiling Finish Material and Application
Kitchen		
Bath		
Other		

Additional information: ______________________________

16. Interior Doors and Trim:

Doors: type ______________ ; material __________ ; thickness ______

Door trim: type ______________ ; material __________ ;

Base: type ______________ ; material __________ ; size ________

Finish: doors ______________ ; trim __________

Other trim *(item, type and location)* ______________________________

Additional information: ______________________________

17. Windows:

Windows: type ________ ; make ________ ; material ______ ; sash thickness ____

Glass: grade __________ ; ☐ sash weights; ☐ balances, type _____ ; head flashing _____

Trim: type ______________ ; material ______________________

Paint ______________ ; number coats ______________________

Weatherstripping: type ____________ ; material ______________________ ;

Storm sash, number ______________________________

Screens: ☐ full; ☐ half; type __________ ; number __________ ; screen cloth material

Basement windows: type ___ ; material __________ ; screens, number ____________________ ;

Storm sash, number ______

Special windows ____

Additional information: ____

18. Entrances and Exterior Detail:

Main entrance door: material ___________ ; width ________________ ; thickness _______".

Frame: material _______________________ ; thickness _____________ "

Other entrance doors: material ___________ ; width ________________ ; thickness _______".

Frame: material _______________________ ; thickness _____________ "

Head flashing _______________________ Weatherstripping: type ___ ; saddles __________

Screen doors: thickness ________________ "; number ______________ screen cloth material

Storm doors: thickness ___________ "; number ______________

Combination storm and screen doors; thickness ___ "; number _______ ; screen cloth material

Shutters: ☐ hinged; ☐ fixed. Railings _____ ; Attic louvers __________

Exterior millwork: grade and species ______________________________

Paint ___________________________ ; number coats ____________________________

Additional information: ____

19. Cabinets and Interior Detail:

Kitchen cabinets, wall units: material _____ ; lineal feet of shelves _____ ; shelf width _______

Base units: material ______________ ; counter top ____________ ; edging ___________

Back and end splash ______________ Finish of cabinets _______ ; number coats _____

Medicine cabinets: make _______________ ; model _____________________________

Other cabinets and built-in furniture ______________________________________

Additional information ____

20. Stairs:

Stair	Treads		Risers		Strings		Handrail		Balusters	
	Material	Thickness	Material	Thickness	Material	Size	Material	Size	Material	Size
Basement										
Main										
Attic										

Disappearing: make and model number ______

Additional information: ______

21. Special Floors and Wainscot:

Wainscot Floors

Location	Material, Color, Border, Sizes, Gage, Etc.	Threshold Material	Wall Base Material	Underfloor Material
Kitchen				
Bath				

Location	Material, Color, Border, Cap, Sizes, Gage, Etc.	Height	Height Over Tub	Height in Showers (From Floor)
Bath				

Bathroom accessories: ☐ Recessed; material ______ ; number ______

☐ Attached; material ______ ; number.

Additional material: ______

22. Plumbing:

Fixture	Number	Location	Make	Mfr's Fixture Identification No.	Size	Color
Sink						
Lavatory						
Water closet						
Bathtub						
Shower over tub △						
Stall shower △						
Laundry trays						

△ Curtain rod △ Door ☐ Shower pan: material ______

Water supply: ☐ public; ☐ community system: ☐ individual (private) system.*

Sewage disposal: ☐ public; ☐ community system; ☐ individual (private) system.*

*Show and describe individual system in complete detail in separate drawings and specifications according to requirements.

House drain (inside): ☐ cast iron; ☐ tile; ☐ other ______

House sewer (outside); ☐ cast iron; ☐ tile; ☐ other ______

Water piping: ☐ galvanized steel; ☐ copper tubing; ☐ other ______ Sill cocks, number __

Domestic water heater: type ____________________ ; make and model ____________________ ;

heating capacity __________ gph. 100° rise.

Storage tank: material ____________________ ; capacity _______ gallons.

Gas service: ☐ utility company; ☐ liq. pet. gas; ☐ other ________________________________

Gas piping: ☐ cooking; ☐ house heating.

Footing drains connected to: ☐ storm sewer; ☐ sanitary sewer; ☐ dry well.

Sump pump; make and model ___________________ capjacity _______ ; discharges into _____

23. Heating:

☐ Hot water. ☐ Steam. ☐ Vapor. ☐ One-pipe system. ☐ Two-pipe system. ☐ Radiators.

☐ Convectors. ☐ Baseboard radiation.

Make and model ____

Radiant panel: ☐ floor; ☐ wall; ☐ ceiling. Panel coil: material ___________________________

☐ Circulator. ☐ Return pump. Make and model ____________________ ; capacity _________ gpm.

Boiler: make and model ____

Output _____________ Btuh; net rating ______________________ Btuh

Additional information: ____

Warm Air: ☐ Gravity. ☐ Forced. Type of system ____

Duct material: supply _____________________ ; return ______________________________

Insulation __________ , thickness _________ ☐ Outside air intake.

Furnace: make and model ____

Input ______________ Bthu.; Output _______________________ Btuh.

Additional information: ____

☐ Space heater; ☐ floor furnace; ☐ wall heater. Input ______________ Btuh.;

Output _____________ Btuh.; number units __________________

Make, model ____

Additional information: ____

Controls: make and types ____

Additional information: ____

Fuel: ☐ Coal; ☐ oil; ☐ gas; ☐ liq. pet. gas; ☐ electric; ☐ other _______ ; storage capacity ____

Additional information: ____

Firing equipment furnished separately: ☐ Gas burner, conversion type. ____________

☐ Stoker: hopper feed ____________ ☐ bin feed

Oil burner: ☐ pressure atomizing; ______ ☐ vaporizing ____________

Make and model ____________ Control ____________

Additional information: ____________

Electric heating system: type ____________

Input ______ watts; @______ volts; Output ____ Btuh.

Additional information: ____________

Ventilating equipment: attic fan, make and model ____________ ; ____________

capacity ________ cfm. ______

kitchen exhaust fan, make and model ____________

Other heating, ventilating, or cooling equipment ____________

24. Electric Wiring:

Service: ☐ overhead; ☐ underground. ______ Panel: ☐ fuse box; ☐ circuit-breaker;

Make ______ AMPs ______ No. circuits ____________

Wiring: ☐ conduit; ☐ armored cable; ☐ nonmetallic cable; ☐ knob and tube; ☐ Other ____

Special outlets: ☐ range; ☐ water heater; ☐ Other ____________

☐ Doorbell ☐ Chimes ☐ Push-button locations ____________

Additional information: ____________

25. Lighting Fixtures:

Total number of fixtures ____________

Total allowance for fixtures, typical installation, $ ____________

Nontypical installation ____________

Additional information: ____________

26. Insulation:

Location	Thickness	Material, Type, and Method of Installation	Vapor Barrier
Roof			
Ceiling			
Wall			
Floor			

Hardware: *(make, material, and finish.)* ____________________

____

____

Special Equipment: *(State material or make, model and quantity. Include only equipment and appliances which are acceptable by local law, custom and applicable FHA standards. Do not include items which, by established custom, are supplied by occupant and removed when he vacates premises or chattels prohibited by law from becoming realty.)* ____________________

____

____

____

____

____

____

____

____

27. Miscellaneous:

(Describe any main dwelling materials, equipment, or construction items not shown elsewhere; or use to provide additional information where the space provided was inadequate. Always reference by item number to correspond to numbering used on this form.) ____________________

____

____

____

____

____

____

____

____

____

Porches: ____

____

____

Terraces: ____

____

____

Garages: ____

____

____

Walks and Driveways:

Driveway: width ______________ ; base material ___ "; thickness ________ "

surfacing material ______________ ; thickness _______ "

Front walk: width ______________ ; material _______ ; thickness ________ ".

Service walk: width ______________ ; material _______ ; thickness ________ ".

Steps: material ______________ ; treads _________ "; risers ___________ ".

Cheek walls ____

Other Onsite Improvements:

(Specify all exterior onsite improvements not described elsewhere, including items such as unusual grading, drainage structures, retaining walls, fence, railings, and accessory structures.)

____

Landscaping, Plainting, and Finish Grading:

Topsoil ______________ " thick: ☐ front yard; ☐ side yards; ☐ rear yard to ___________

feet behind main building.

Lawns *(seeded, sodded or sprigged):* ☐ front yard ______________ ; ☐ side yards _____ ;

☐ rear yard ____

Planting: ☐ as specified and shown on drawings; ☐ as follows:

Shade trees, deciduous, ______________ " caliper.

Low flowing trees, deciduous, ____________ ' to ____________ '

High-growing shrubs, deciduous, ____________ ' to ____________ '

Medium-growing shrubs, deciduous, __________ ' to ______________ '

Low-growing shrubs, deciduous, ____________ ' to ______________ '

Evergreen trees, __________________________ ' to ______________ ', B&B.

Evergreen shrubs, _________________________ ' to ______________ ', B&B.

Vines, 2-year ____

Identification. — This exhibit shall be identified by the signature of the builder, or sponsor, and/or the proposed mortgagor if the latter is known at the time of application.

Date ____________________ Signature ____

____________________ Signature ____

FHA Form 2005

VA Form 26-1852

Certificate of Insurance

ALLSTATE INSURANCE COMPANY HOME OFFICE — NORTHBROOK, ILLINOIS

Name and Address of Party to Whom this Certificate is Issued

Name and Address of Insured

INSURANCE IN FORCE

Type of Insurance and Hazards	Policy Forms	Limits of Liability	Policy Number	Expiration Date
Workmen's Compensation		STATUTORY*		
Employers' Liability	Standard	$ PER ACCIDENT (Employer's Liability only) *Applies only in following state(s):		
Automobile Liability		Bodily Injury / Each / Property Damage		
☐ owned only	☐ Basic	$ PERSON		
☐ non-owned only	☐ Comprehensive	$ ACCIDENT $		
☐ hired only	☐ Garage	$ OCCURRENCE $ Bodily Inj. and Prop. Dam. (Single Limit)		
☐ owned, non-owned and hired		$ EACH ACCIDENT $ EACH OCCURRENCE		
General Liability		Bodily Injury / Property Damage		
☐ Premises—O.L.&T.	☐ Schedule	$ EACH PERSON		
☐ Operations—M.&C.		$ EACH ACCIDENT $		
☐ Elevator	☐ Comprehensive	$ EACH OCCURRENCE $		
☐ Products/Completed Operations		$ AGGREG. PROD. COMP. OPTNS. $		
		AGGREGATE OPERATIONS $		
☐ Protective (Independent Contractors)	☐ Special Multi-Peril	AGGREGATE PROTECTIVE $		
		AGGREGATE CONTRACTUAL $		
☐ Endorsed to cover contract between insured and __________ __________ __________ date__________		Bodily Inj. and Prop. Dam. (Single Limit) $ EACH ACCIDENT $ EACH OCCURRENCE $ AGGREGATE		

The policies identified above by number are in force on the date indicated below. With respect to a number entered underpolicy number, the type of insurance shown at its left is in force, but only with respect to such of the hazards, and under such policy forms, for which an "X" is entered, subject, however, to all the terms of the policy having reference thereto. The limits of liability for such insurance are only as shown above. This Certificate of Insurance neither affirmatively nor negatively amends, extends, nor alters the coverage afforded by the policy or policies numbered in this Certificate.

In the event of reduction of coverage or cancellation of said policies, the Allstate Insurance Company will make all reasonable effort to send notices of such reduction or cancellation to the certificate holder at the address shown above.

This certificate is issued as a matter of information only and confers no rights upon the certificate holder.

Date ____________________ , 19 _______________ By ___________________________________

Authorized Representative

U454-16

(8-81)

Reading the Plans

On the following pages are a set of plans, with each plan covering pages and drawn to serve a specific purpose.

For example, the plan on pages 122 and 123 is labeled "Foundation Plan" at the lower left corner. It shows all of the concrete work associated with this house, including the walls, piers, foundation for fireplace, and the slabs for the garage and the patio.

The plan on pages 124 and 125 is the floor plan. Study yours, and ask yourself specific questions about living in a house that is built to your plan. Will you be happy with only a sliding glass door in the family room? Do you want to provide space for both a dining room, for formal dining, and a dinette, or would you prefer to make other use of one of those spaces? Do you want your garage doors opening to the rear of the house, or closer to the front entrance?

Look at the detail sheet on the next two pages. It shows the second floor. Have you always wanted a fireplace in your master bedroom? It probably could be arranged — at extra cost. Will you be happy with the laundry up there (where most dirty clothes end up) or would you prefer it on the first floor?

Study the details on pages 128–129. Note the drawing on the left. It shows 3½ inches of insulation — the width of a common 2×4. Are you satisfied with that, or do you live in a cold climate where double that amount of insulation would pay for itself in lower fuel bills? Is the kitchen layout what you want, or are there points of it that will become irritations when it is used?

And finally, the outside elevations. Are you happy with its looks? Do you want that outside chimney, or would you prefer to see it inside, where its stored heat would be fed back into the home? Will you feel proud of this house when you drive up to it — or when your friends do?

A thorough study of the plans should give you the feel of living in the house. You'll know where your TV set will be placed — and your library of books. You'll have a place for the holiday decorations, the garden tools, the tools and equipment of your hobbies. Your furniture will fit in the spaces planned for it.

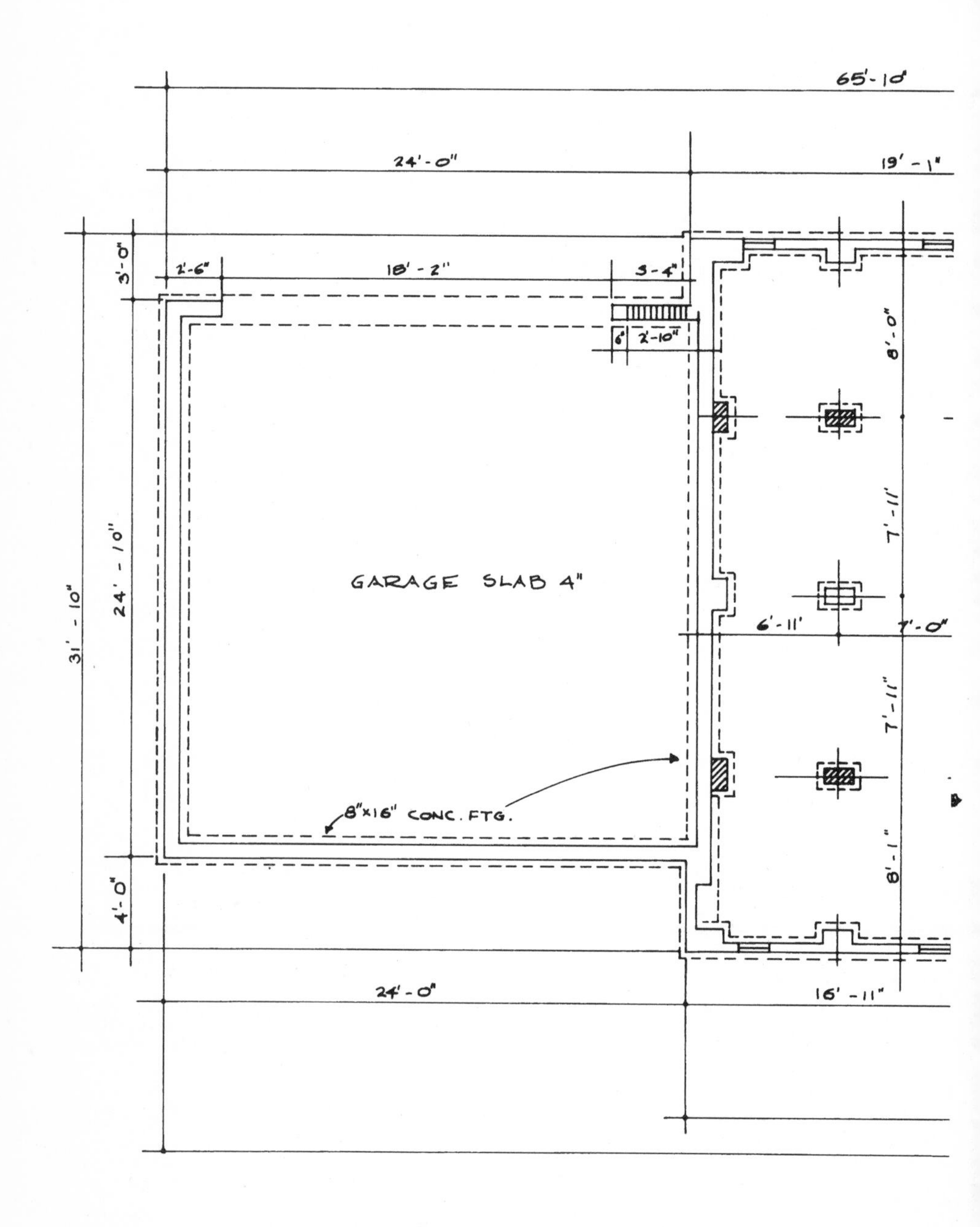

Foundation Plan

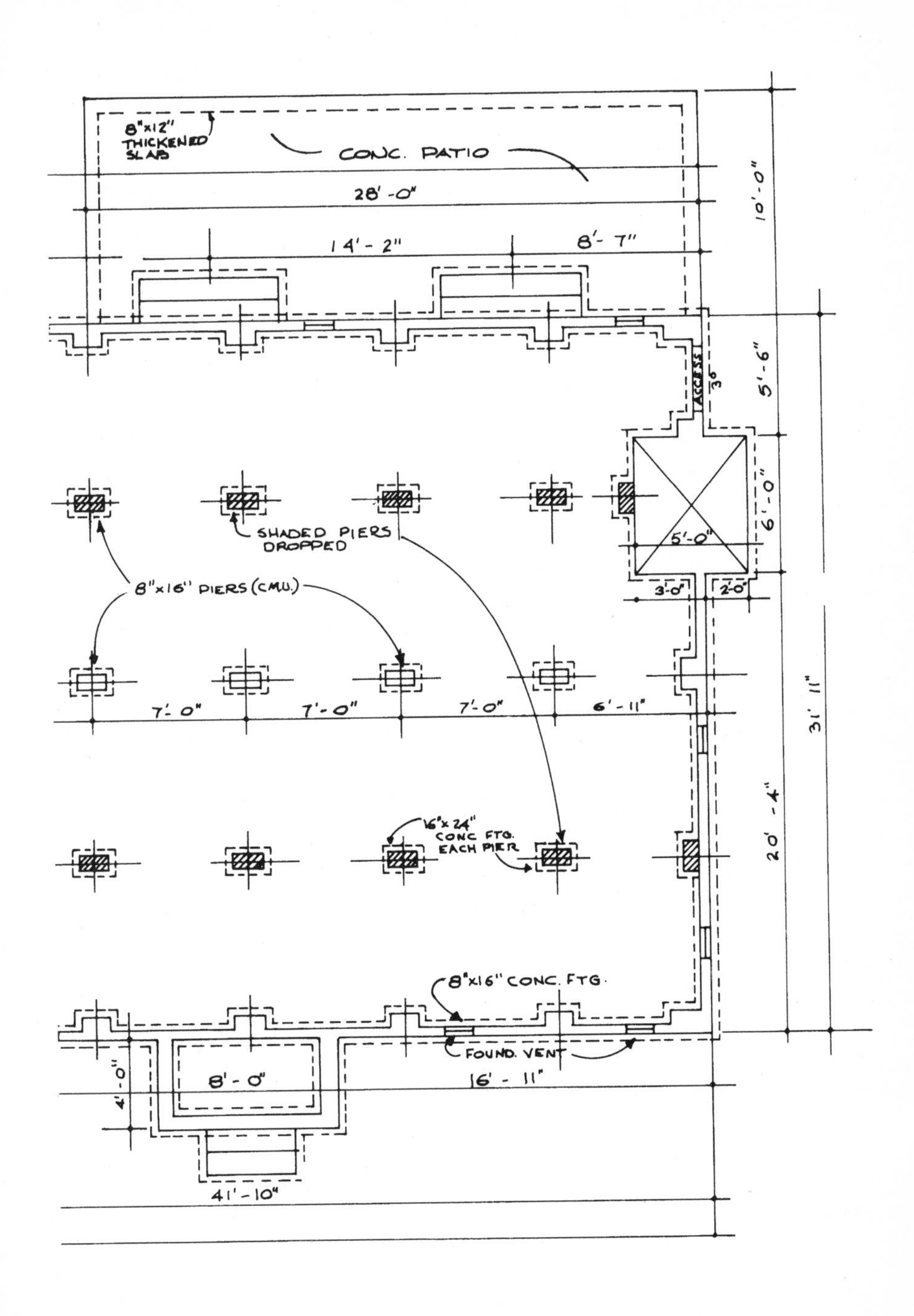
8"x12" THICKENED SLAB
CONC. PATIO
28'-0"
10'-0"
14'-2"
8'-7"
ACCESS
36"
5'-6"
6'-0"
5'-0"
3'-0"
2'-0"
SHADED PIERS DROPPED
8"x16" PIERS (CMU)
7'-0"
7'-0"
7'-0"
6'-11"
31'-11"
16"x24" CONC FTG. EACH PIER
20'-4"
8"x16" CONC. FTG.
FOUND. VENT
4'-0"
8'-0"
16'-11"
41'-10"

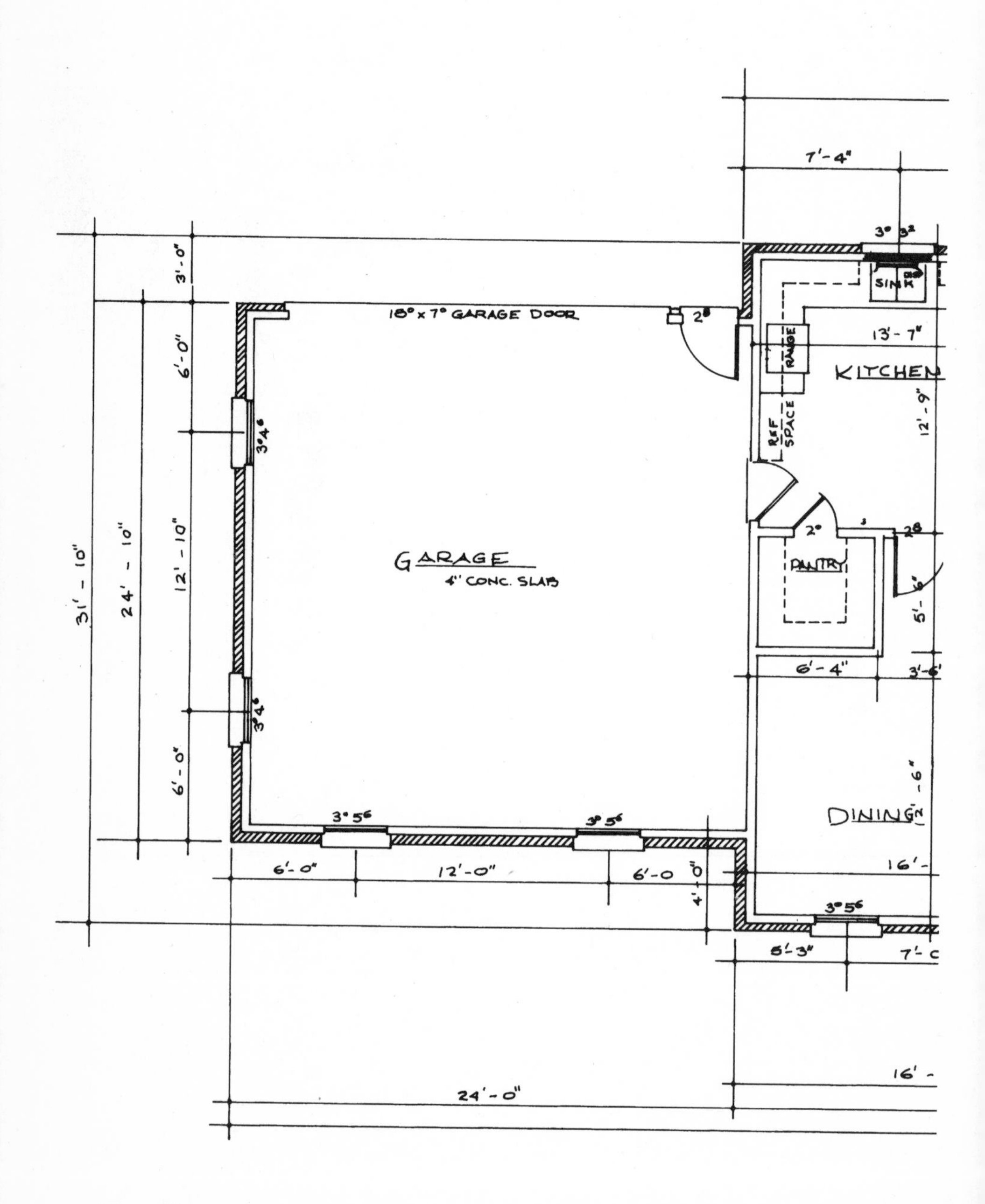
18° x 7° GARAGE DOOR
GARAGE
4" CONC. SLAB
KITCHEN
SINK
RANGE
REF SPACE
PANTRY
DINING
7'-4"
3'-0"
6'-0"
12'-10"
24'-10"
31'-10"
6'-0"
6'-0"
12'-0"
6'-0
4'-0"
24'-0"
13'-7"
12'-9"
6'-4"
5'-3"
16'-

Floor Plan

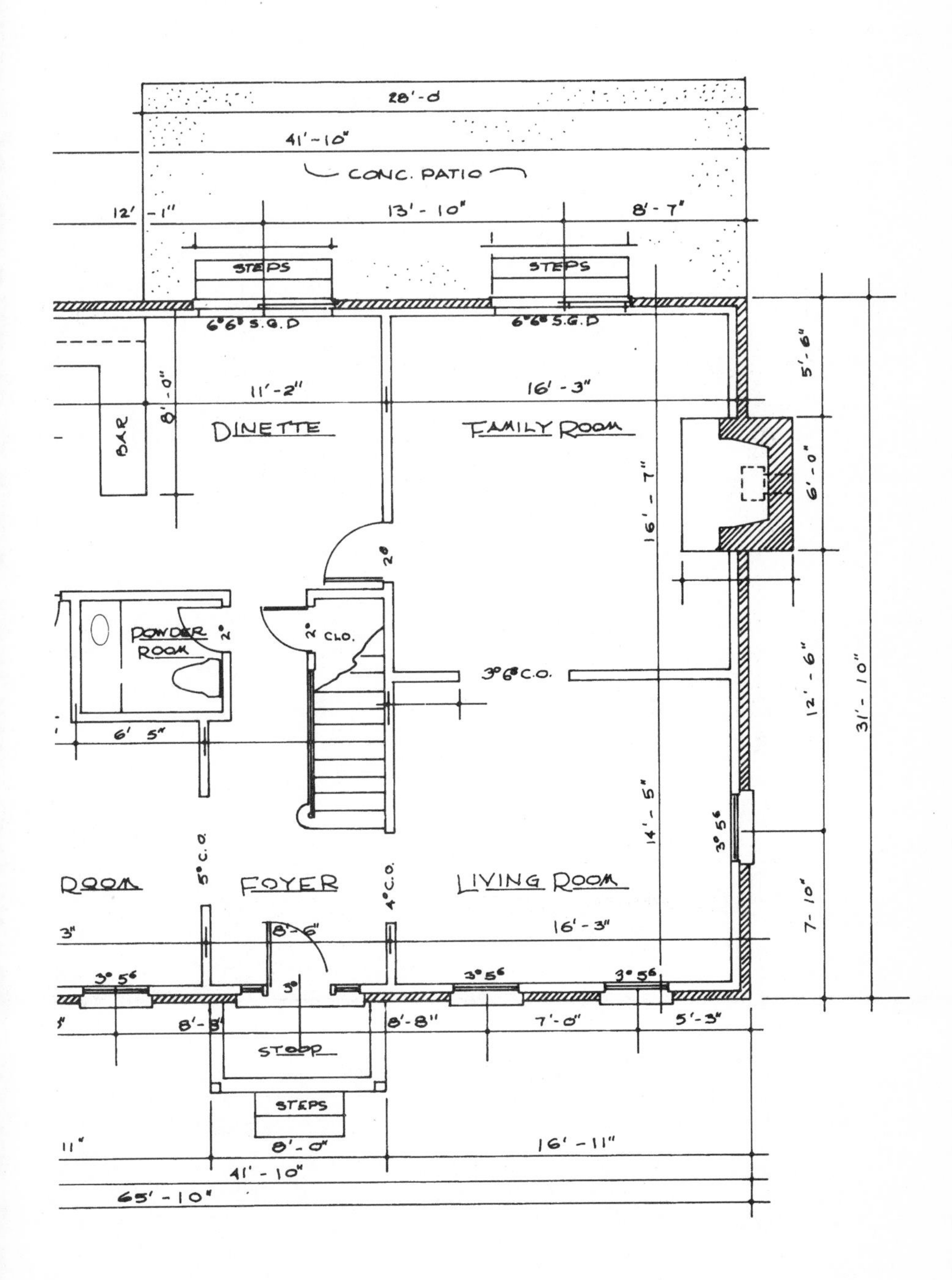
28'-0
41'-10"
CONC. PATIO
12'-1"
13'-10"
8'-7"
STEPS
STEPS
6°6° S.G.D
6°6° S.G.D
11'-2"
16'-3"
8'-0"
BAR
DINETTE
FAMILY ROOM
5'-6"
6'-0"
16'-7"
2°
POWDER ROOM
2°
2°
CLO.
3°6° C.O.
12'-6"
31'-10"
6' 5"
14'-5"
3° 5°
5° C.O.
4° C.O.
ROOM
FOYER
LIVING ROOM
7-10"
3"
8'-6"
16'-3"
3° 5°
3°
3° 5°
3° 5°
8'-3"
8'-8"
7'-0"
5'-3"
STOOP
STEPS
11"
8'-0"
16'-11"
41'-10"
65'-10"

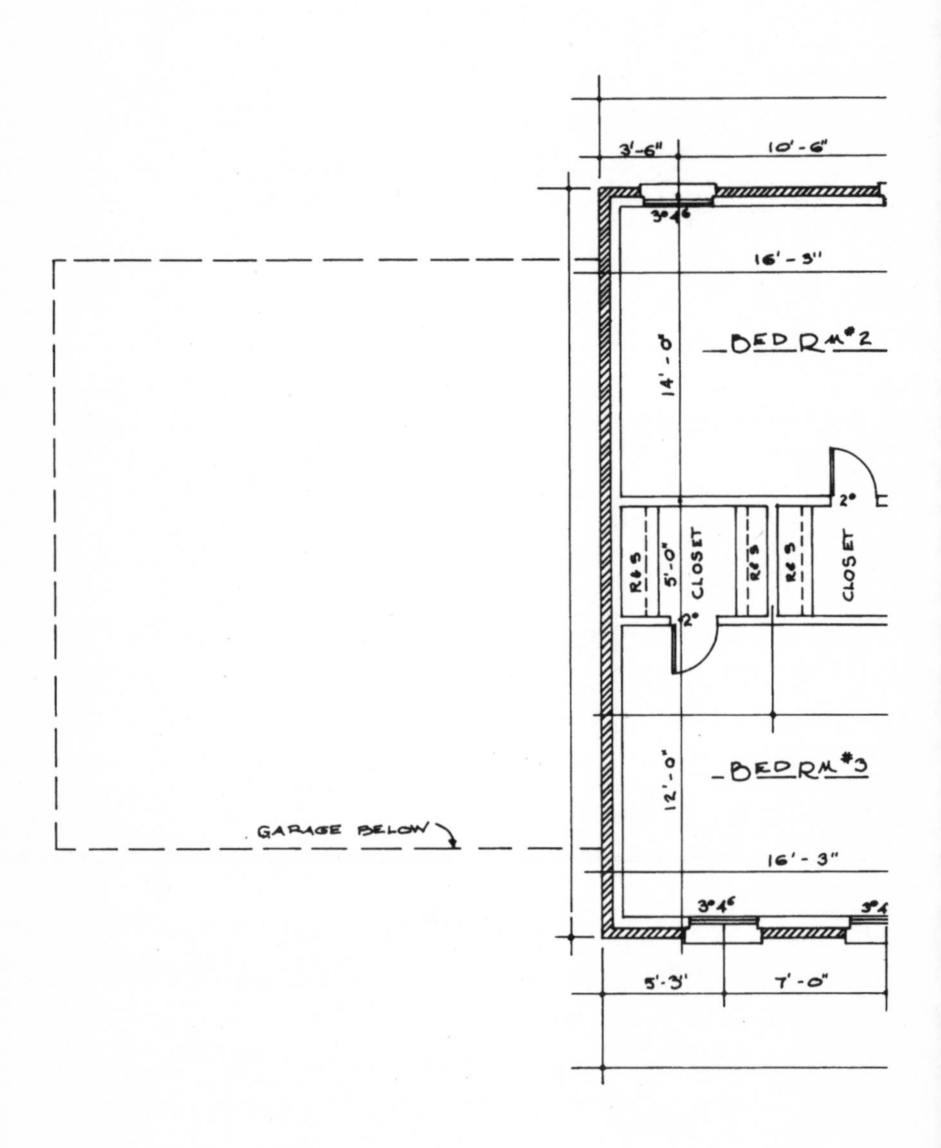

Detail Sheet

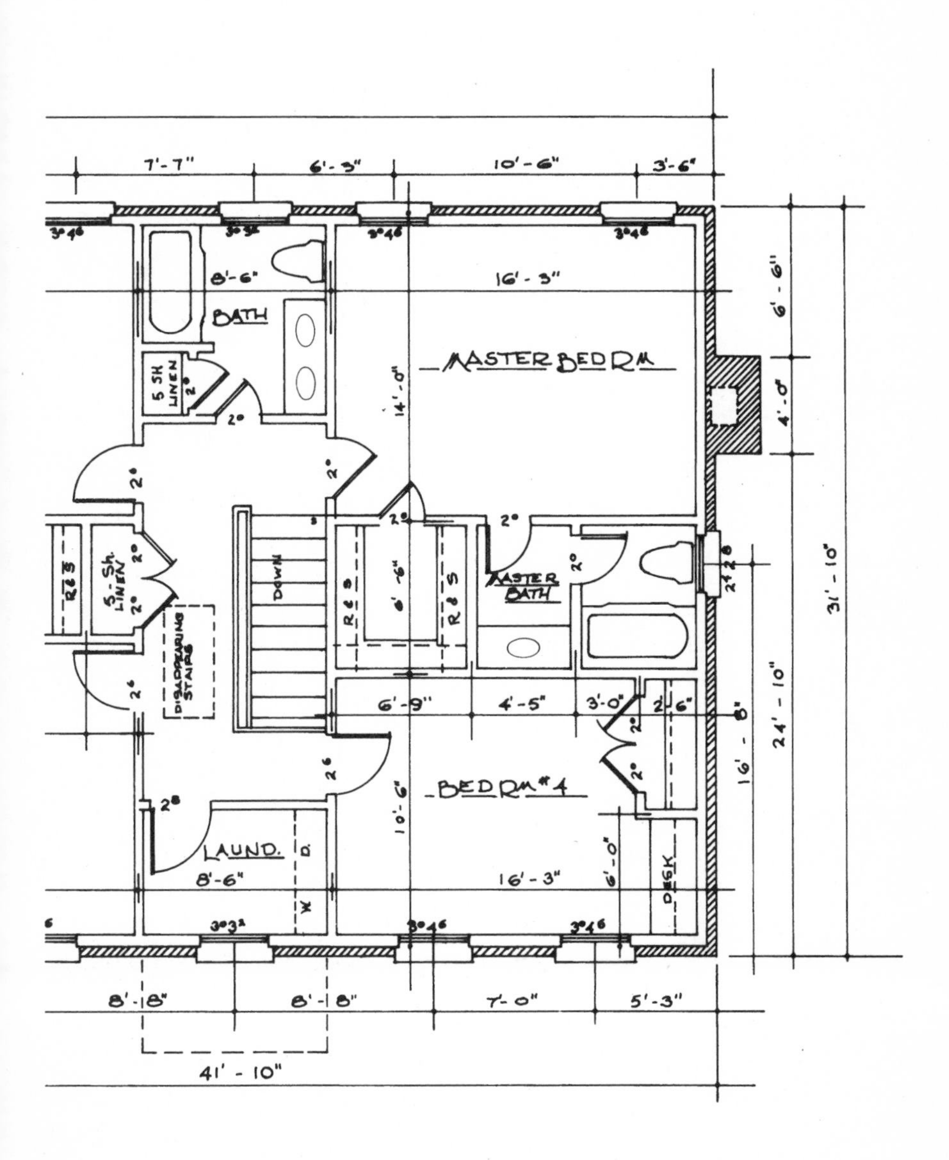
7'-7"
6'-3"
10'-6"
3'-6"
8'-6"
BATH
16'-3"
MASTER BEDRM
14'-0"
6'-6"
4'-0"
5 SH LINEN
DOWN
MASTER BATH
DISAPPEARING STAIRS
6'-9"
4'-5"
3'-0"
BEDRM #4
10'-6"
LAUND.
8'-6"
16'-3"
6'-0"
DESK
31'-10"
24'-10"
16'-8"
8'-8"
8'-8"
7'-0"
5'-3"
41'-10"

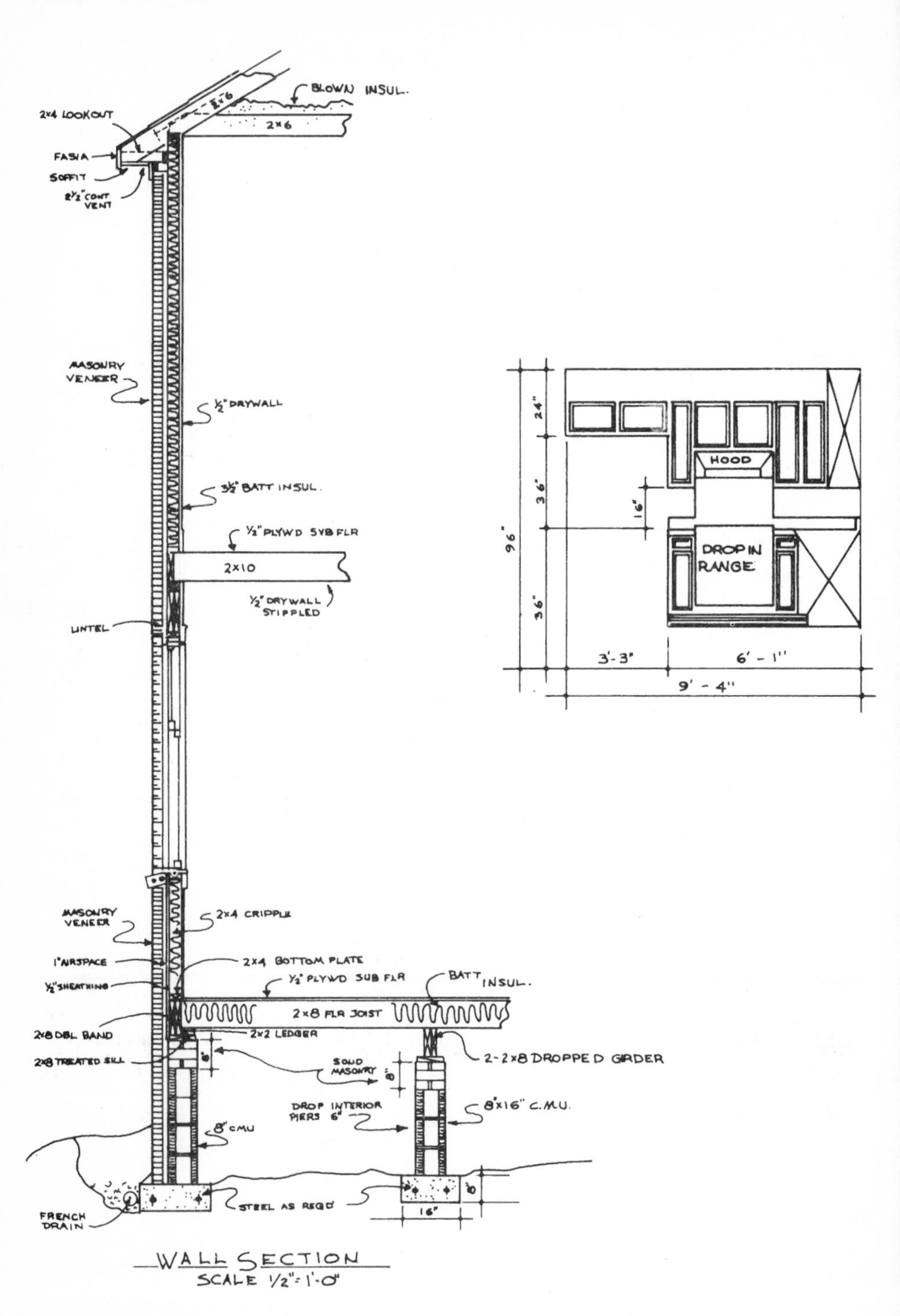

Wall Section

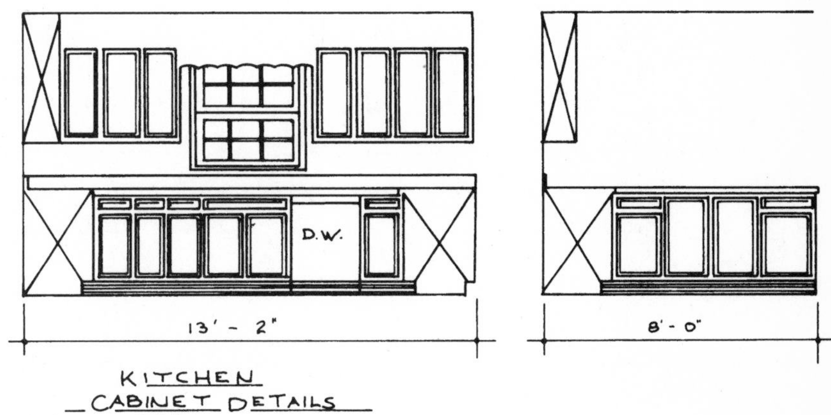

KITCHEN
CABINET DETAILS
3/8" = 1- 0"

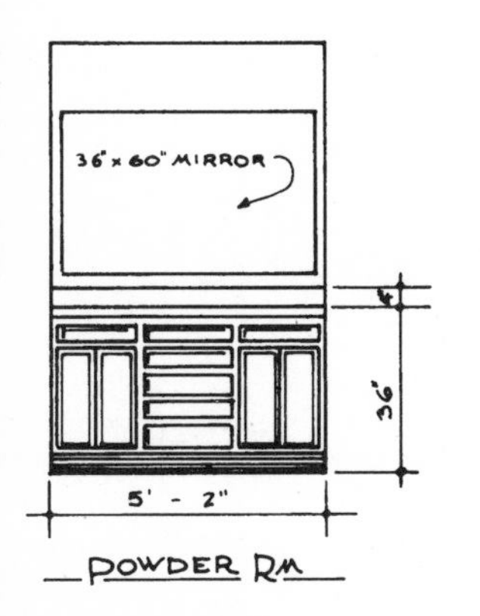

POWDER RM

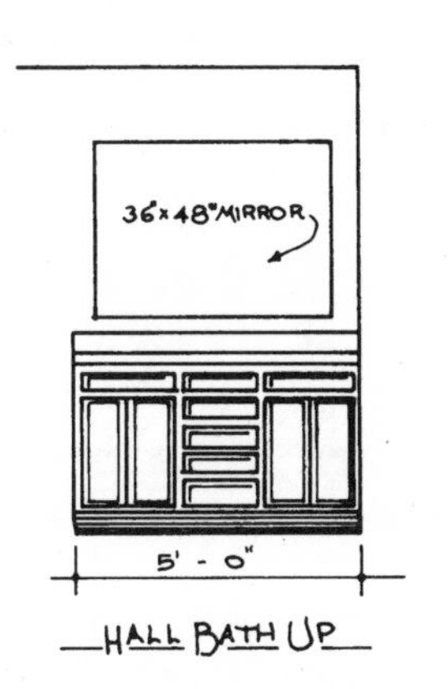

HALL BATH UP

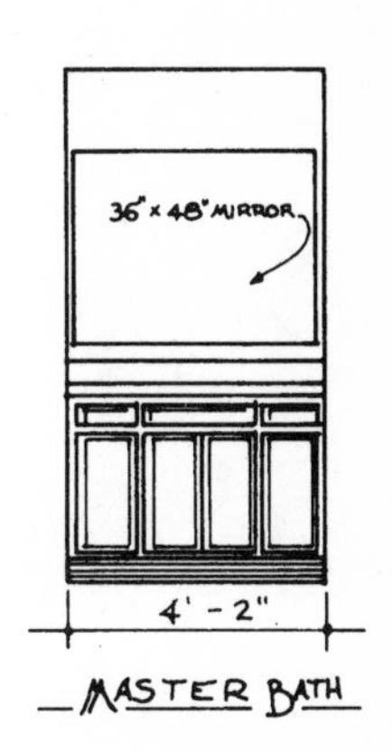

MASTER BATH

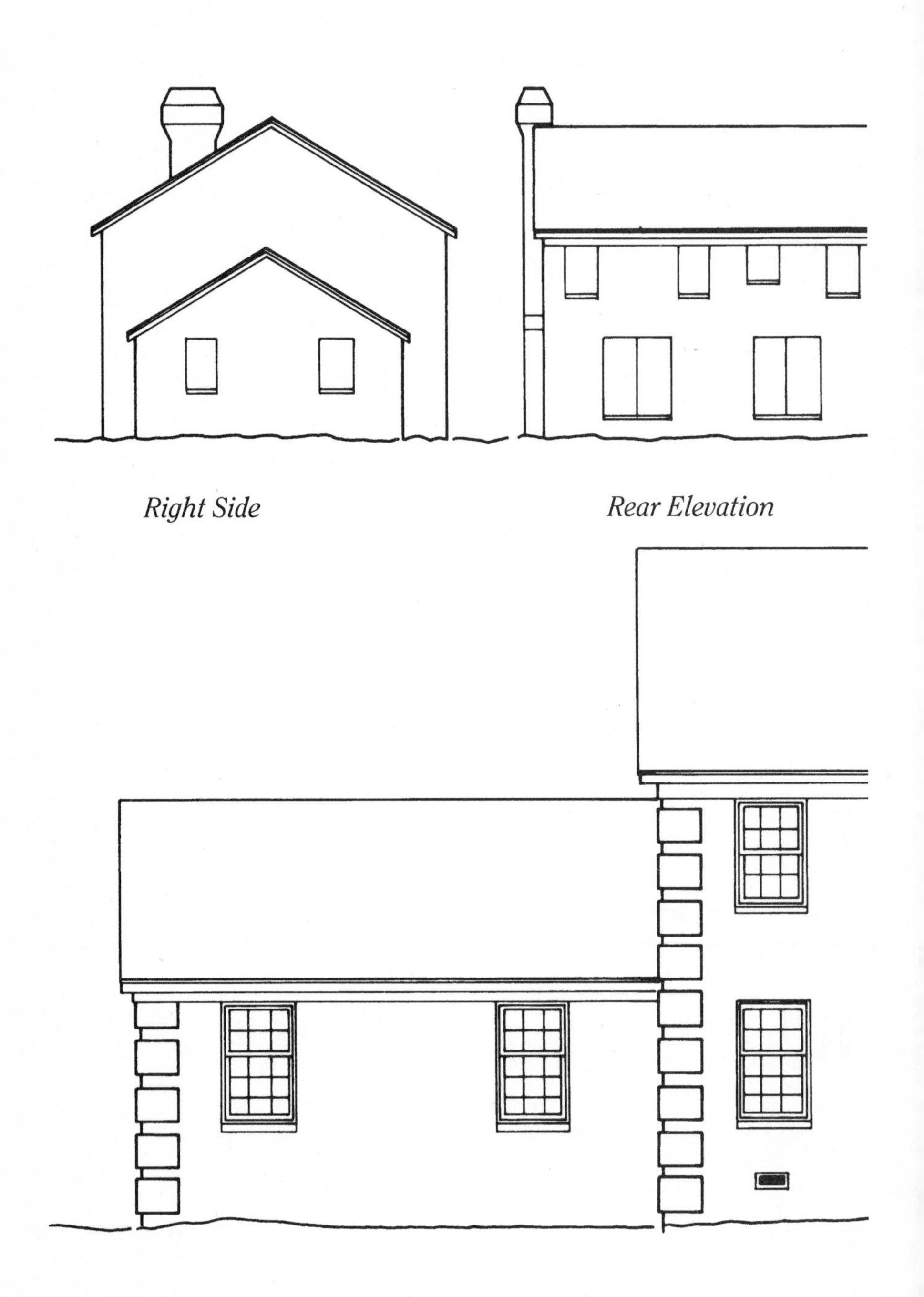

Outside Elevations

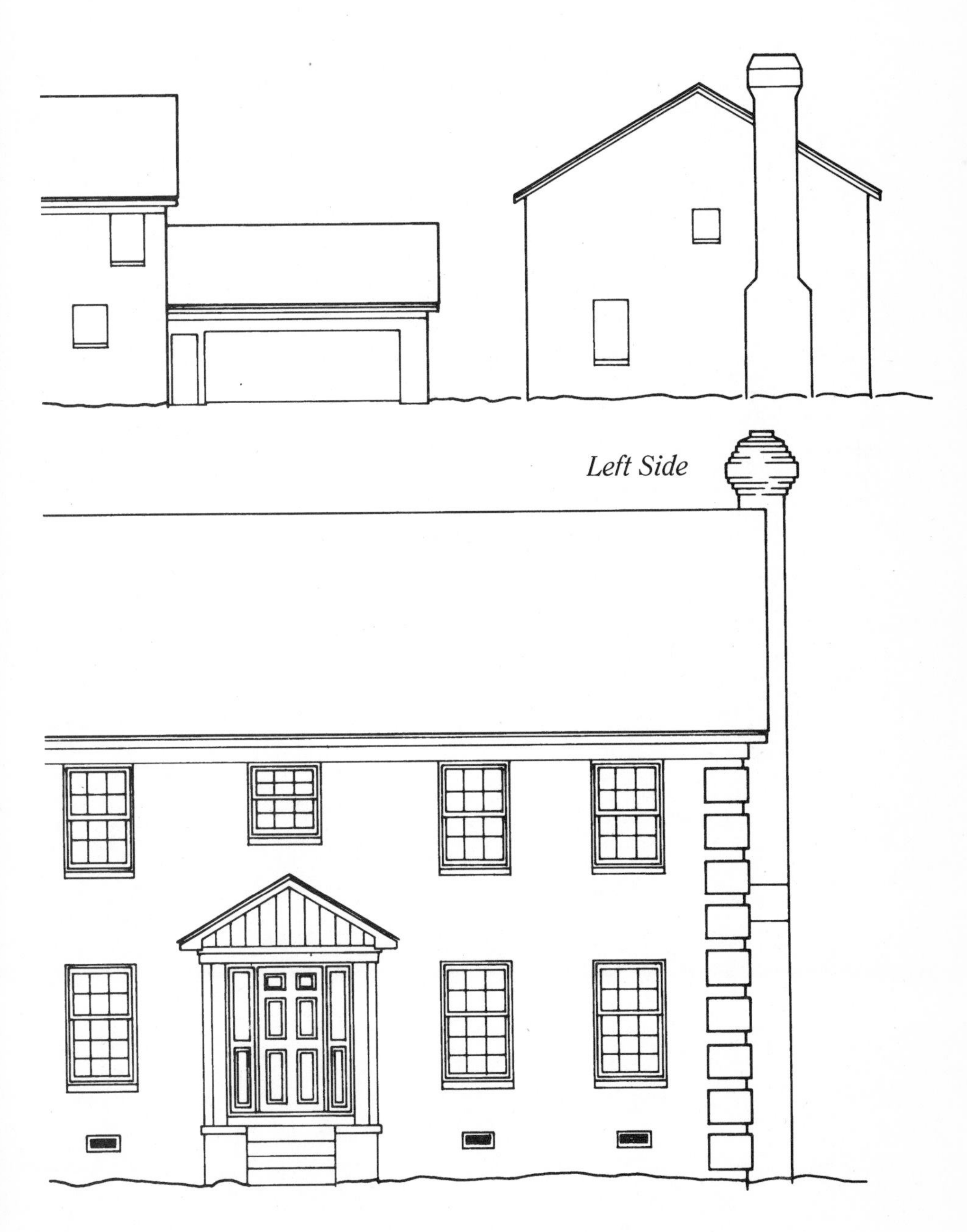

Left Side

Front Elevation

Index

Page numbers in *italics* indicate illustrations

H

I

J

L

M

T

U

V

W

Z